Evo-illusion
of Man

By Dr. Stephen T. Blume'

Dedication

Evo-illusion of Man is dedicated to the best teacher I ever had: Mrs. Maizlish. I wish I knew her first name, but I never addressed her any other way. Mrs. Maizlish was a very unpopular mathematics teacher at my high school, Hollywood High. In the days when I was at Hollywood High, there were no computers to help us with our scheduling. We made up our class schedules by running for classes when the bell rang at the top of the hour. It was a virtual full panic stampede every hour on the hour. We students would make a list of the most desirous teachers, which usually meant the easiest. If our first choice class was filled, then we would run for the next on our lists... then the next. In my senior year, every advanced math class on my list, except the last, was filled when I arrived. My very last choice was Mrs. Maizlish. The people who didn't get classes, as was the case with me, slowly sauntered in to Mrs. Maizlish's class. There was no hurry, as she was the strictest and most unpopular math teacher in the school. She was last on everyone's list. When I walked into her class, I just sank. She was sitting on top of her front desk, not in her desk chair; with a slight grin, as if she knew exactly what was happening. I can remember this scene as if it were yesterday. She was about a 50-year-old Jewish lady, a bit overweight, with a rather large hooked nose. She looked a bit like the Wicked Witch of the North. She didn't look friendly; or easy.

I sat down at an open desk, and watched as other students filed in slowly, tired from their sprint to the other filled math classes. They had the same bad luck as I did; and the same look on their faces. They were out of breath and grief-stricken because they had just signed up for the hardest and strictest teacher in the school.

It turned out that Mrs. Maizlish was the most wonderful, and memorable teacher I ever had in my life. Mrs. Maizlish taught me how to think; I mean how to *really* think. She taught me how to break complex problems down into their components, and to solve them one component at a time; how to turn complex word problems into paragraphs that told the story of the problem. A story was always far easier to solve and make an equation for than was a problem. She would stand over me and encourage me, and tell me I *am* intelligent enough to solve *any* complex math problem. "Now Stephen, write the problem like a paragraph. Let X = the number of gallons, then X-36 equals..." She made me believe in myself.

When I was a senior in dental school I made plans to walk into her class at Hollywood High, put my arm around her in front of the students, and plant a big kiss on her cheek; and say thank you so much. I called the school to see when a good time would be. I was shocked to find out Mrs. Maizlish had died of a heart attack two weeks before my call. I was so sad. I berated myself. Why did I wait so long to thank her? Mrs. Maizlish changed my life, and she had no idea. I cannot think of a better person to dedicate this book to than you. This is my second chance to say thank you Mrs. Maizlish, and I am taking it. I wish you were here to accept it.

PREFACE

My dad, Carl Blumay, early in his career, was a sportswriter for the Los Angeles Times. He later took a post with Occidental Petroleum. He was able to see the company rise from a $50,000 loser to a $19,000,000,000 oil giant during the twenty-five years he was there. He wrote a book about his experience at Oxy, and with its CEO, Dr. Armand Hammer. It was titled, *The Dark Side of Power*. It was published by Simon and Schuster. His first draft was over 1,000 pages long. Simon and Schuster liked his first draft, but they wanted it cut down to fewer than 300 pages, which is the magic number for maximum book sales. The rewrite took him several more years of work. By the time he made it to press, Armand Hammer, the lead character of the book, had been deceased for a couple of years. The delay significantly attenuated my dad's book sales. He did sell nearly 100,000 copies, but it would have been many more if he had originally written much less and released it in a timelier manner. In this situation, less was certainly more.

I do love writing so very much. I am certain I inherited the writer's bug from my dad. I am so fascinated with the sciences that make us humans tick, and that are involved in our origins, like microbiology, biochemistry, biology, anthropology, paleontology, genetics... My dad didn't have that fascination at all. I don't know whom I inherited that interest from. I'm kind of a black sheep in my family.

As I wrote my first book, *Evo-Illusion*, I surprisingly completed more pages than Simon and Schuster's magic number of 300 pages before I discussed human evolution. So, I had to save the evolution of man for this, my second book. Lucky me. I got to write another book! And this one was sure a fun write. When I started, I had a pretty good notion about what human evolution was all about. I was pretty familiar with the human phylogenetic tree, and the incredible cast of characters. But as much as I thought I knew about human evolution, I still got a great education; and a shocking and amazing insight about its *modus operandi*.

Since writing *Evo-illusion*, I thought if there were any part of evolution science that might trump what I had to say in *Evo-Illusion*, the evolution of man would be it. The fossils were so well preserved, so locked in. They were present in every major museum, including the Smithsonian National Museum of Natural History, and the Field Museum in Chicago, where I first became cynical about evolution after being a very devoted fan. There was a really nice and believable stream of hominids that lead to humans. People who I discussed evolution with would usually bring up, "What about all of those early humans!" It was a tough argument to counter. Hominids and their timelines seemed so real; so valid. I decided I would write in the direction that this book and its research would take me. If it looked like humans evolved from an earlier ape ancestor, I would write that. If there were holes in the theory, I would write about that. The conclusions of this book were to be formed by my findings, and nothing else.

This book takes a detailed look at all of those early pre-humans. It's a far more interesting subject than I thought possible. Really digging in, and seeing what human evolution is composed of was a real eye-opener; a very intriguing undertaking.

The research for this book *was* fascinating. I gained a whole new perspective on what modern science can do, and actually is. I found that the myths and tales of science from ancient times have followed us into the 21st century. We modern humans aren't immune and cured of concocting fables any more than the ancients were. We know so much more now than the ancients, but the gaps in our knowledge and our lack of answers in certain fields are still filled in with illusions and fables, just like theirs were. In many ways we are no different than they were. They simply had much larger gaps to fill, and miniscule scientific knowledge to fill them with, so their myths were much larger and more obvious. This book will relay the story of human evolution, and the scientific illusions that we ss have concocted to fill our gaps of knowledge, just as the ancients did.

I have been accused of having an agenda; of wanting to prove that the source of all of living nature is or was some sort of religious god, and that religion is part of my agenda. Nothing could be farther from the truth. My only agenda in writing *Evo-Illusion* and *Evo-Illusion of Man* is the promotion of good science, and the fun it brings. Every time I sit down at the computer to add to this book, I have a very euphoric feeling. The fun of the write will far exceed whatever financial successes or failures I may have with *Evo-Illusion of Man*. I hope you enjoy the read as much as I enjoyed the write. I hope it will be the same adventure for you as it was for me.

In Evo-illusion of Man I use a few terms that I coined. One is *evo-illusion*, the definition of which is pretty obvious. The conjurers of evo-illusions are *evo-illusionists*. These are lecturers, writers, and speakers that promote evo-illusions. The other term is *audience*. The audience is made up of students that sit in on evo-illusionist's lecturers, people who watch TV documentaries promoting evo-illusions, and read books on the subject. They are the people who are constantly being fooled by evo-illusionists. I was an enthusiastic part of the audience for a very long time.

Evo-illusion of Man

Table of Contents

Note: This is the current definition of hominids and hominins:

Hominids – all modern and extinct great apes. Gorillas, chimps, orangs and humans, and their immediate ancestors. Not gibbons.

Hominin – Any species of early human that is more closely related to humans than chimpanzees, including modern humans themselves.

In the past the term *hominid* referred to a species that was in the process of evolving from ape to human. If you Google the term hominid and go to images, all you will see are pre-human species. The definition of the term hominid has been modified to include humans with apes and primates. This change benefits the notion that all humans are apes, which in turn benefits the teaching of evolution of man. For the purposes of this book, when I write the term *hominid*, I am discussing the sub-human species that supposedly evolved into humans. Species like *Australopithecus afarensis* and *Homo habilis*, and the like. It does not include modern humans.

Chapter 1

Is It An Ape, or Is It An Ape-Man?

Science cannot solve the ultimate mystery of nature because, in the last analysis, we ourselves are part of the mystery that we are trying to solve- Max Planck

In my first book, *Evo-Illusion*, I discussed evolution and how it could not be the source of all living organisms. I discussed what might have happened when the first cells formed, and how evolution says they morphed, through millions of changes and millions of years, into whales and birds. I did not discuss what evolution's scientists say about how an ape common ancestor evolved into modern apes and humans. Human evolution, if it truly occurred, is a subject unto itself. This book is dedicated to taking a critical look at that subject. Human evolution is really what it's all about. If it occurred, humans evolved from early apes hundreds of times faster than any other form of evolution. Human

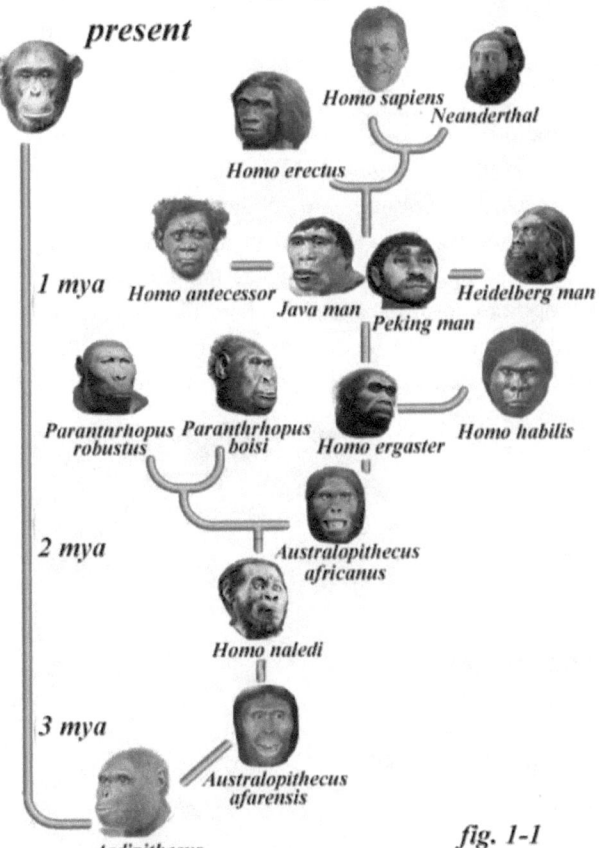

present

Homo sapiens
Neanderthal
Homo erectus

1 mya Homo antecessor
Java man
Peking man
Heidelberg man

Parantnrhopus robustus Paranthrhopus boisi Homo ergaster Homo habilis

2 mya Australopithecus africanus

Homo naledi

3 mya Australopithecus afarensis

Ardipithecus ramidus

fig. 1-1

evolution produced human intelligence and consciousness, the only time this occurred out of millions of opportunities over millions of years in millions of other species. Why is human evolution so unique? The source of *us* is why we're so fascinated with this subject. What could be more important than nature's *modus operandi* for producing humans on Earth?

Fig.1-2

There are two diagrams that do a great job of promoting the evolution of man from ape. They seem so logical, and they make so much sense, until you take a real good look. Figure 1-1 is a *family tree* diagram that details all of the different fossil pre-humans and shows where they fit on the tree. Yes, that's me at the top. Figure 1-2 is a chart that shows the imaginary steps early apes took whilst evolving into modern humans. These diagrams are in most anthropology books and displayed in most anthropology classrooms on Earth. When I was majoring in biology at the University of Southern California, my anthropology professor had a chart like Figure 1-2 mounted at the front of the lecture hall. When he described evolution, and how all of living nature came together because of gradual changes that were naturally selected, and how modern man evolved from apes over millions of years because those changes were beneficial, it made so much sense. I was an immediate and dedicated believer in evolution for many years. Looking at Figures 1-1 and 1-2 alone will sell many people on the illusion that humans evolved from apes. These are such great illusions; they are easily accepted as evidence certain. These two diagrams quickly sold me on the evolution of man. Finally, the origin of mankind made sense. When I saw these two diagrams, and when my professor confidently lectured on how we evolved

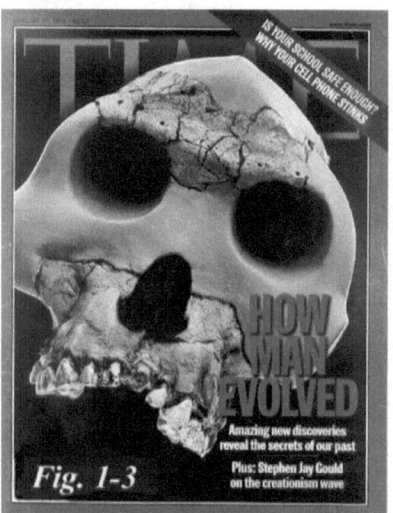

Fig. 1-3

from apes over millions of years, I was sold. These charts look so real, so scientific. It's hard to look at them and not believe what they have to say. Are they real; or are they illusions? Is the foundation of human evolution valid?

I do realize that the vast majority of scientists in the world are strong believers that humans arose through ape-to-human evolution. It's simply a universally accepted concept. Even Time Magazine (Figure 1-3) from August of 1999 tells us "how man evolved". They never ask the biggest question: **Did** man evolve from an ape common ancestor? How could little ole' me prove that the world's scientists and Time are flat out wrong? My job in this book is to show you that human evolution is nothing but a concocted illusion. You will know for yourself once you see the evidence, drawn purely from the facts.

To understand ape to human evolution, you need a basic understanding of the clear differences between the bones of apes and humans. They are easy to spot. Taking a few moments learning about the differences will make it far easier for you to make a scientific assessment of the validity of human evolution. If I can make you an expert in just a few minutes of reading, you will be able to decide for yourself what is real and what is not. You won't have to believe and accept what someone else tells you. If you are an avid believer in human evolution, just keep an open mind. The information itself will tell the tale. You won't have to believe the pronouncements of any human, including me. You can see for yourself.

There are many characteristics that could be cited in the full skeleton of humans and apes that would determine to which group a particular skeleton belongs. I am going to cite only their skulls, hands and feet. Each *individual* characteristic that I cite will separate apes from humans. I am exampling a chimpanzee because they're genetically the closest primates to humans. Evo-illusionists claim chimpanzees have 96 percent of human DNA. If we have a common ancestor with chimps, they are our cousins many times removed. So

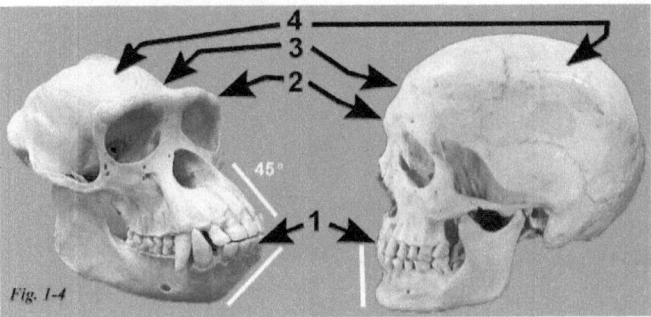
Fig. 1-4

our skulls should be similar in design as well. But there are a few obvious and very defining characteristics that need to be known if one is to determine if ancient ape bones are just from apes; or if they truly represent our ancestors.

Figure 1-4 is a montage of the skulls of a chimpanzee (left) and a human (right). The following are key characteristics of ape skulls that human skulls do not have:

(1) **Prognathic Jaws**: A chimpanzee's maxilla (upper jaw) and mandible (lower jaw) protrude significantly. Typically, the bone from the nose to the tip of upper teeth extrudes out at about a 45^0 angle, whereas a line drawn from the nose to the chin of a typical human is vertical or concave.

(2) **Large Brow Ridges** above the eye sockets. Humans have negligible brow ridges.

(3) **Absent or small laid back forehead**: Apes lack significant vertical foreheads. Humans have large vertical foreheads, which provide room for the much larger *frontal lobe* of our brains. This is an important difference, as the large frontal lobe of the human brain allows us the ability to make decisions and solve problems. It also controls our behaviors, voluntary movements, emotions, and consciousness. Without a forehead, an animal would not have room for a large frontal lobe, and could not perform functions that differentiate animals from humans. The ability to make tools, improve on them, and the ability to remember how to make them, wouldn't be possible without a forehead and large frontal lobe.

(4) **Small Ovoid or Flat Cranium** that houses their much smaller brains. Human adult craniums are about two to three times the volume of ape craniums. Ape craniums are narrower than the lateral extents of the eye sockets, whilst human craniums are far wider than the outer extent of their eye sockets. (See Figure 1-7 below)

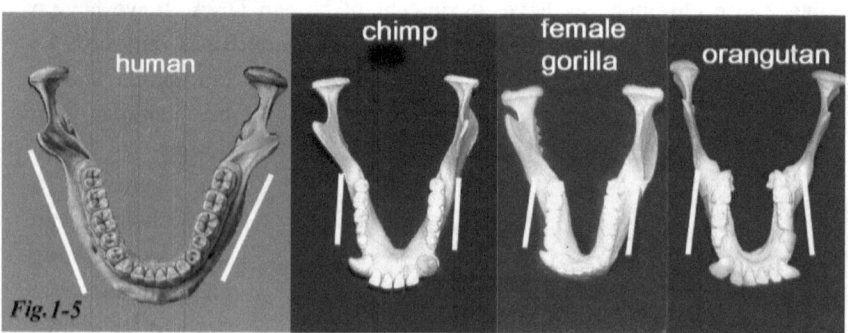

Fig. 1-5

(5) **Posterior Dental Arches Are "U" Shaped Or Even Towed In Toward The Back**: (Figure 1-5) Human dental arches are shaped like a rounded "**V**". The lineup of human posterior teeth is flared toward the back. If you look at the chimp, gorilla, and orangutan dental arches you will see that the right and left posterior (back) teeth are lined up parallel to each other. (white lines) The orangutan's teeth actually flare to the front.

(6) **Apes Have a Shorter and Lower Positioned Thumb than Humans:** (right image in Figure 1-6) Notice, the thumb doesn't reach the first knuckle of the index finger. Compare with your own hand.

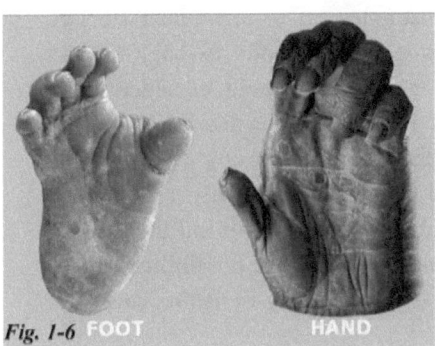

Fig. 1-6 FOOT   HAND

(7) **Ape Big Toes Are Separated From The Rest Of The Toes, and are aligned 65⁰ away from the other toes.** (left image in Figure 1-6) Ape feet almost look like hands. Ape feet and hands are designed so they each act as both hands *and* feet. Their hands and feet are both proficient at walking *and* grasping. Our hands are graspers, not walkers, whilst our feet are walkers not graspers.

There you go. Now you're a trained expert at telling the difference between ape fossils and human fossils. Each of these characteristics taken individually can separate ape from man. For example, a skull with large brow ridges will always be ape, not human. Brow ridges alone are a defining characteristic.   Now that you're an expert, let's give you a test. Take a look at Figure 1-7 and see if you can tell which skulls are ape, and which are human. Hint: *(A)* is hominid. Or is it ape? Observe how the cranium of a human is wider than its eye sockets, whilst the smaller ape craniums terminate inside of the eye sockets. (arrows)[1]

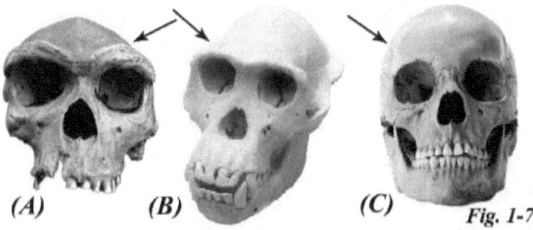

(A)   *(B)*   *(C)*   Fig. 1-7

When diggers unearth supposed ancient pre-human fossils, all possibilities must be considered, not just a single possibility that is given by evo-illusionists with an agenda who are trying so hard to form their illusions. The fact is evo-illusionists routinely use *the* single possibility that supports their illusions to describe fossil finds when other possibilities exist and are far more likely. This is a demonstration of what an incredible illusion human evolution is. So, what are the possibilities that exist when an ancient ape skeleton, or any part thereof, no matter how tiny, is unearthed by diggers and paleoanthropologists? Evo-illusionists assign them to be human precursors and place them on a branch of the human tree of life. They must do this when there are so few fossils, or the illusion of human evolution will crash. They can't choose if a particular fossil is an ape or sub-human because there is such a dearth of samples. Evo-illusionists declare nearly every ape-like bone and tooth dug out of the dirt to be a human precursor. Are there other possibilities that scientist should consider when fossils of supposed sub-humans are found? Fortunately for this book and me there are. Here are a few of the possibilities that should be on any *true* scientists mind when analyzing supposed early ape-person fossils. Only (7) will be considered by any evo-illusionists for certain. The first six will kill the illusion so you will never hear of them from any evo-illusionists. The possibilities are:

(1) It's a modern ape skeleton that was damaged by moisture, pressure, and chemical decomposition.

(2) Ninety-nine percent of all species that have ever inhabited the Earth became extinct. This means a large number of primate (ape, monkey) species became extinct as well. If a conservative percentage of primates became extinct, say 25 percent over the last five or ten million years, that would mean there are over 200 extinct primate species. These extinct primates certainly died and left bones all over Africa. The odds are huge that "hominid" fossils represent nothing but extinct ancient apes.[2,3]

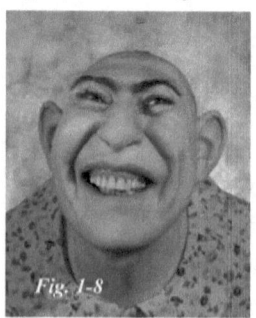

Fig. 1-8

(3) It's a conglomeration of bones found in separate locations put together by evo-illusionists to make them appear to be from a single hominid. In other words, it's a hoax. This has a much higher chance of being the case than one would think, as you will see.

(4) It's a bone or part of a skeleton of an ancient non-primate vertebrate species that looks similar to an early ape.

(5) They're ancient bones of a true human that were crushed, broken, or disturbed in a way that makes them seem a like a human precursor.

(6) They're the bones of a *microcephalic* human: or a human who had major deficiencies and mental deficits and were intellectually underdeveloped. Figure 1-8 is the photo of a micro cephalic person with a slanted forehead and small cranium. Microcephaly occurs in about one in seven thousand births. Microcephalics are generally incapable of inventing and making tools or any complex devices. Almost every hominid fossil found that is supposedly an early human is either an ape whose braincase is "micro" compared to humans, or a microcephalic human. Since no vertebrate with a micro-brain is capable of making tools, any tools found near a microcephalic fossil could not have been invented and made by the live animal that left the fossil.[4-6]

(7) They're hominids, or pre-human. They evolved over hundreds of thousands or millions of years into modern man.

Of all the seven possibilities, (7) is by far the least likely of the choices, and the least provable. But it's the most often used to describe puny and dubious fossil finds by evo-illusionists; as I will demonstrate.

For the purposes of this book, I would like to pose a law. This law seems beyond obvious, but it's never considered or discussed by any evo-illusionists I have ever come across, with good reason. It kills their illusion. So here it is:

Blume's Law of Hominids:

If a selected individual feature of an early fossil skull or skeleton of an ancient animal is similar to an equivalent feature on a human skull or skeleton, that is not evidence whatsoever that the fossil species evolved into humans.

Everyone seems to agree, by default, that if any feature of an ancient skull has any single human-like feature, even though dozens of other features are ape or animal features, it is assumed that that one feature trumps all others, and the fossil is a pre-human ape that evolved into Homo sapiens. Evo-illusionists will frequently cite the shape of a single tooth on an ape skull that is similar to a tooth on modern humans to help form their illusion that hominids evolved into humans. In doing so, they ignore and distract their audience from noting numerous other major characteristics that show just the opposite. There is zero evidence that any early ape fossil was from a species that evolved into humans. Even if it did happen, it cannot be proved. This fact is ignored and glossed over in the world of evo-illusion. For almost two centuries it's been a "given" that early apes and supposed hominids evolved into humans; it's an accepted edict. In reality, this is just another fascinating scientific illusion. But in this illusion it's a given that most people accept without question.

One of the biggest scientific illusions involving the evolution of man is the equating of brain volume with intelligence. Evo-illusionists cite the brain size of different hominids to support the illusion that will make you think they were in the process of evolving into humans. "Craniums that were 800 cc, evolved into 900 cc craniums, which lead to 1400 cc human craniums!" Anatomic parts of the brain, for example a significant frontal lobe, tells a lot more about the abilities of an animal than brain size. Brain volume can be an indicator of intelligence, but it cannot be completely reliable. If the brain size theory were valid, elephants and whales would be the smartest species on the planet. On average, elephant brains are about 7,500 cubic centimeters in volume. Sperm whale brains are about 7,200 cubic centimeters. Human brains average about 1,400 cubic centimeters. If brain size is a key indicator of intelligence, elephants and whales should be inventing computers and space shuttles. Just imagine an elephant named Neil Armstrong being the first on the moon. Wouldn't that be a hoot? Referring to an increase in brain size as evidence of ape to human evolution is part of the illusion. Modern primates have varying brain volumes. One thing is for certain. No matter what the brain volume is of any of the 625 modern primate species, humans are the only ones that can make a tool. [7,8]

One would think that if evolution were valid, there would be dozens of modern apes in the process of evolving foreheads and larger brains so they might become toolmakers and cave artists, just as pre-humans did. But apes score a perfect 100 percent. Not one primate species, except for humans, has a significant vertical forehead and large brain, or is in the process of evolving them. Anyone with the tiny bit of knowledge about bones proffered earlier in this chapter should be able to easily determine if the once-living animal source of an unearthed skull was able to make cave drawings and tools. If a tool was found next to an ancient unearthed fossil skull, and if that skull had no forehead, the finders first thought should be to wonder where the hell that tool came from. It sure couldn't be from the animal that owned that skull. The step that is usually taken is to declare the tools were made by the animal source of the skull. In reality, there is no possible way to link an unearthed fossil bone to rock tools that were found nearby. They cannot be tied together.

What is very interesting about the illusion of human evolution is the numerous unearthed alleged pre-human fossils that supposedly evolved into conscious intelligent humans over comparatively short periods of time; in hundreds of thousands to maybe a couple of million years. When looking at the fossils of every other animal, many exist in the fossil record for hundreds of millions of years without change. Nautilus has been in the fossil record for 500 million years without change. Frogs for 350 million years without change. Sharks for 450 million years. The list is long of species that have existed for eons without change. They then either became extinct, like the dinosaurs, or they exist as modern species; like frogs, nautilus, and sharks. If you want evidence for *stasis*, the non-changing of species over eons, go to any museum of natural history with a good fossil collection. The lack of change in species will be obvious; and astounding. Most prominent evo-illusionists have noted, recognized, or written about stasis. Stasis isn't just my observation; it's recognized by all of biology, paleontology, and evolution science as well. Of course they have a dubious explanation called *punctuated equilibrium* that I discussed in *Evo-Illusion*.[9]

Unlike the stasis in both the plant and animal fossil records, human-to-ape evolution had to produce new species, new entities, new characteristics, and finally us, in comparatively diminutive bits of time. Why would our species produce so much evolution in such short time spans when millions of other species didn't? Frogs were frogs, nautilus was nautilus, and sharks were sharks for hundreds of millions of years. But humans gained their entire set of unique characteristic differences in short bursts of less than a few hundred thousand years. What's so special about apes and humans? Why did we suddenly become such rapid evolvers? The possibility with the greatest chance of being valid is that we didn't evolve rapidly from early apes; that ape to human evolution is nothing but an illusion believed by many.

When I was a fan of evolution, I always thought the series of fossils of human ancestors were the most secure evidence of all that evolution is how we humans came to be. By inference, if humans evolved from apes in the manner the fossil record shows, all animal species, in fact all living species of any kind, had to arise in the same fashion as humans. Human fossils and human precursors were undoubtedly overwhelming evidence for evolution. How could anyone doubt? I certainly didn't, until my fateful visit to the Field Museum in 2001 where I saw no evolution in the fossil record. That trip was the stimulus for my wondering, questioning, and researching. But always lingering as an argument in favor of evolution were "those guys": Cro-Magnon, Java man, Peking man, Lucy… They were certainly a force to be reckoned with. They surely were evidence that evolution was the answer to the greatest *Puzzle* of all time.

NOVA and PBS made a documentary on human evolution in 2011 titled *Becoming Human: Unearthing Our Earliest Ancestors*. The announcer described rocks that were found at a dig that were chipped in a particular way, which made them seem like tools made by intelligent pre-humans. Apes gradually increased in intelligence, through naturally selected mutations, until they were able to make these stone tools. Supposedly these stone tools were the first sign of intelligence in human precursors. The fact that these were chipped in a certain way so they could be used as tools for shaping and cutting proved intelligence in their design. Of course, as time went on, our ancestors evolved more and more complex brains, with greater and greater intelligence, until we became intelligent enough to make cars, airplanes, and computers.

The announcer said:

In 2.5 million-year-old layers, scientists begin to find something new. We might be tempted to call them rocks. **But someone was shaping them!** *They are the first stone tools.* **The way we know this is a tool instead of just a broken rock is that it's broken in a very particular way.** *Like it's broken this way, that way, this way... back and forth. There was a method behind the way this rock was broken to make it into a tool. And it's not a random method.*

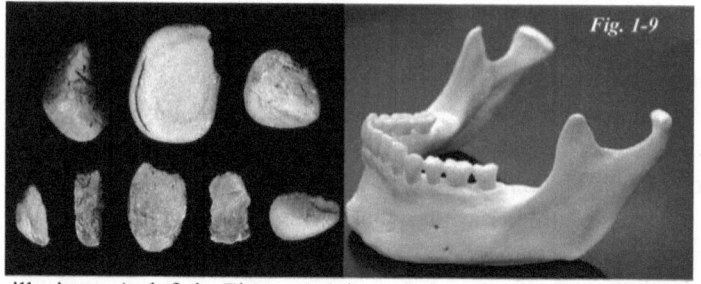

Fig. 1-9

I find this to be the most perfect example of why evolution is nothing but an incredible illusion. At left in Figure 1-9 is a photo of rocks that NOVA thinks are stone tools. NOVA says these are a great example of the beginnings of human intelligence. I'm sorry, but I don't think I'd have any trouble finding rocks like those in any rock pile. The right photo is a human mandible; a lower jaw. It has two condyles designed like the balls of two ball and socket joints that fit perfectly into its human skull socket counterpart. Both joints are equipped with engineered cushioning. Each condyle slides forward and back, independently, which makes the mandible a *double sliding double ball and socket joint*. The mandible has teeth that, if orthodontically straight, mesh-like machine-made parts with the teeth in the maxilla (upper jaw). It has small linear canals, or tunnels, that carry nerves and blood vessels. The little "dots" on the side are openings for the sensory and motor (muscle) nerves of the lower face. They are also exits for blood vessels that feed the surrounding tissues. The mandible is an ingenious device, but the announcer and NOVA would like you to believe there was not a lick of intelligence necessary in its design. But those chipped rock "tools"? Oh boy, it took oodles of intelligence to design those rock-tools. In fact, the rock tools are proof of the beginning of the evolution of human intelligence. The jaw? Even though it's infinitely more complex, far more useful, and ingeniously designed, no intelligence was needed or used to invent and construct it. Compare and contrast the two. Which requires more intelligence for its manufacture? The answer is obvious. See how evo-illusions work? The audience can be so easily fooled into thinking the rocks are intelligently designed, whilst the jaw isn't.

The NOVA documentary displayed the ground, ostensibly around where a

Homo erectus fossil was discovered. (Figure 1-10) The announcer said:

Here is a cache of 500 hand axes made by Homo erectus. A NOVA paleontologist says: *Under my feet are thousands of fragments of stones from toolmaking.* He then picks up broken rocks that look like... broken rocks. He claims these are failed attempts at making stone tools.

Fig. 1-10

So viewers must assume those rocks *were* hand axes, and if the illusion works, they will. Of course, I have questions. Did these "stone tools" sit on the ground for hundreds of thousands of years, untouched and unmoved? Why

would Homo erectus die right where there are hundreds of hand axes just sitting around? Was each of these hundreds of stone tools individually unearthed? Is it possible the stone tools were made comparatively recently by modern humans? Are they really stone tools at all? Is there any way to make sure those stone tools truly

Fig. 1-11

were sculpted by Homo erectus, when, over hundreds of thousands of years, the possibility of sources for those stone tools is vast?

The viewers are then shown a "hand axe" that ostensibly was made by Homo erectus. (Figure 1-11) This is classic illusion technique. You are shown a close-up of a well-engineered hand axe, then a picture of a bunch of rocks on dirt. Most viewers will surmise that the rocks on the dirt are the same excellent tools as the well-designed and well-constructed hand axe. Except NOVA blew this illusion. The hand axe had sharp and fresh looking cutting edges and well-defined chips made during its sculpting. It appeared to be recently constructed. There is no possible way a crisply constructed and engineered hand axe could have those sharp edges after sitting on the ground for hundreds of thousands of years. Wind, sand, and water erosion would have rounded all of those nice sharp edges. NOVA presents a perfect example of a scientific illusion. If the audience can be fooled by the illusion of one well-done prop, they can then be easily fooled by dozens of other props, even if they aren't so flawless. The NOVA host continued:

But there are other clues to his (Homo erectus) *intelligence. The stone tools he left behind. Homo erectus made tools like this hand axe that's been chipped extensively on both sides. The point allows it to do piercing tasks. It can be used for cracking bone or chopping wood. It's a very very versatile tool, and a sharp one. It may not look like much, but the stone hand axe marks the birth of technology. Homo erectus has left us many signs of his inventiveness. The kind of decision-making it takes to make a stone tool has been researched extensively by John Shay.*

John Shay proceeds to make a hand axe just like the hand axe supposedly found near and credited to the Homo erectus fossil. *I'm just going to tap it a little bit. I'm just checking it out to see if there are any internal flaws.*

John finishes his hand axe. Strangely it looks exactly like the hundreds of thousands of years old "sample" ostensibly made by Homo erectus. Is it possible John Shay made the supposed Homo erectus hand axe? Maybe he did, and accidentally dropped it in that field of stone tools, he forgot, and it got kinda mixed up with them. Then someone accidentally picked it up and gave it to the NOVA director, and he said, "Hey... look what I found in that bunch of rocks!"

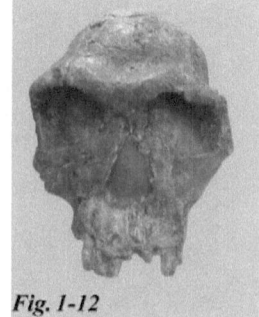
Fig. 1-12

The announcer continued: *Even for an expert, making a hand axe is not easy. A good toolmaker has to understand the properties of stone. Many of these stones have hidden defects, and failing to spot them could spell disaster… The skilled craftsman, Homo erectus, had developed a new type of intelligence.*

Figure 1-12 is a Homo erectus skull. He is obviously devoid of a forehead, and therefore, he is also devoid of a significant frontal brain lobe. He has a flat skullcap, and tiny cranium. There are zero primates on the Earth that have these features that are capable of being "skilled craftsmen". The announcer noted that even for a human expert, the making of a hand axe would not be easy. Does this guy look like he would understand the properties of stone? Could he chip and chop a hand axe out of a chunk of rock? Could he be an expert? The answer is certain. There isn't a chance in hell this guy made a hand axe or any stone tool. In this evo-illusion he did, and he has fooled many audiences so far into falling for the illusion. Will the audiences ever figure out they have been fooled? If at least a few dozen of the modern 625 primate species on Earth had the intelligence to make stone tools, and to create cave writings and art, and if they could master other skills that required intelligence as should be the case if evolution were truly the source of all living creatures, then NOVA would have a case. But since 100% of all modern primate species with the minuscule cranium and lack of forehead demonstrated by Homo erectus fossils cannot make stone tools, it must be presumed that Homo erectus didn't make rock tools; and NOVA has presented an illusion. [10]

So, if the ape-person who was buried near the hundreds of rock tools was not the maker of those tools, what are the possibilities regarding their source? Here is my list. I'm sure there are more possibilities:

(1) The rock "tools" could have been formulated by the digging crew and placed nearby to make it look like the source of the tools was the ape-person represented by the fossil. Things like this have been done numerous times by over-eager fossil hunters who want to exacerbate their finds. Just think of how much more valuable a supposed hominid fossil would be if tools were found nearby. Nearby tools can create the illusion that makes an ordinary ape fossil appear to be an ape-person fossil. Tools make an ape fossil thousands of times more valuable. The financial and fame incentives are overwhelming.

(2) They're rocks that have been chipped by another group of people who populated the area anytime over hundreds of thousands of years. How can it be proved that the single ape-person that represented a fossil find was the actual

maker of the chipped rocks found "nearby"? There are no fingerprints on those stones.

(3) They are not tools at all, but simply rocks that were chipped and shaped by geological events over millions of years that made them look like tools. Out of the thousands of broken rocks that can be found near any dig, carefully selecting ones that look like tools would not be a difficult undertaking. According to evo-illusionists who say selected random happenstance can make a human skeleton, random geological happenstance, and careful selection by diggers could certainly produce chipped rocks that look a lot like tools. Think of the incentive for the workers on a dig to "find" stone tools and make their bosses very happy.

I always wonder why tools were discarded right where a hominid fell dead or was killed and was buried. Why would the burying committee leave small rocks, chipped into sharp edges, with the dead? I've read papers where it was stated "thousands of tools were found near an unearthed hominid." Thousands of tools around a single dead ape-person? Didn't the hominid population have further use for those "tools"? Did these guys die suddenly when they were sitting around the campfire with their tools in hand, or when they were making those tools? Did a hominid tool-guy that was making those thousands of tools croak from *tool-making exhaustion syndrome*? Tools sure weren't easy for these early apes to make. Why would they be discarded with or near the dead? This scenario is simply too transparent and convenient.[11,12]

Darwin published his second major book on evolution, *The Descent of Man, and Selection in Relation to Sex* in 1871. In it, Darwin applied evolutionary theory to human evolution, citing *sexual selection*, or the way we humans choose our mates for procreation, as the modus operandi for human evolution. Animals selectively kill and consume each other, which is the major driver of animal selection, while sexual selection is the major driver of human evolution. *The Descent of Man* discusses evolutionary psychology,

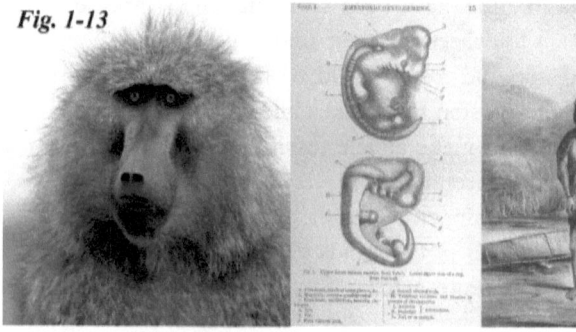

Fig. 1-13

evolutionary ethics, the differences between human races, the differences between sexes, the dominant role of women in choosing mating partners, and the relevance of the evolutionary theory to society. In the introduction to *Descent*, Darwin laid out the purpose of his book:

The sole object of this work is to consider, firstly, whether man, like every other species, is descended from some pre-existing form; secondly, the manner of his development; and thirdly, the value of the differences between the so-called races of man.

Darwin's methodology for maintaining humans evolved like all other animals was to cite how similar human beings are to other modern vertebrates and mammals. He utilized their anatomical similarities. He focused on body structure, embryology, and "rudimentary organs" that he thought were useful in humanity's "pre-existing" forms. Sadly for Darwin, he had no fossil record to prove his theory. There were no rudimentary organs or pre-existing forms. So he had to utilize modern species and their similarities as his evidence. Baboons, dogs and human fetuses, and "savages" provided his chief support for human evolution. (Figure 1-13) He noted the similarity between dog and human fetuses. (Figure 1-13 middle) His encounters with the backward natives of the Tierra del Fuego on his Beagle voyage (Figure 1-13, right) enhanced Darwin's belief that civilization had evolved over time from a more primitive state, just like all animals. [13]

If Charles Darwin's book, On *The Origin of Species*, were considered to be the Old Testament of Evolution, the New Testament would be Richard Dawkins' book *The Blind Watchmaker*. Dawkins is considered to be an evo-Pope and the world's leading evangelist for evo-illusion. The biological sciences have advanced unbelievably since Darwin wrote *Origin*. Dawkins attempted to update Darwin's theory in *Watchmaker*. In it, he said:

...one of the fastest evolutionary changes (is) the swelling of the human skull from an Australopithecus-like ancestor, with a brain volume of about 500 cubic centimetres to the modern Homo sapiens's average brain volume of about 1,400 cc. (It's fast, but really) only .01 cc. per generation. [14,15]

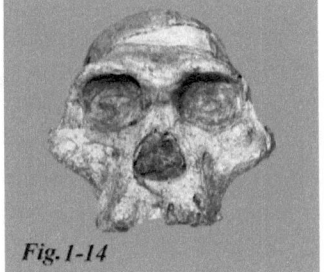

Fig. 1-14

In writing the above statement, did Richard consider that maybe Australopithecus didn't evolve into Homo sapiens? Maybe they were simply separate but somewhat similar species. Wouldn't a good scientist consider this possibility? A good one would, but evo-illusionists like Dawkins don't. Instead they thrive on steering their audiences to fall for illogical illusions by getting people to think they're seeing unfounded and impossible events. The fact is that there are absolutely zero fossils that show that increase in braincase size, as you will see. Dawkins doesn't even consider if the cranial size change from Australopithecus (Figure 1-14) to Homo sapiens was represented by centimeters, the evolution from Australopithecus to human consciousness, cognitive thinking, human emotions, verbal communication, and all the other inconceivable features of the human brain would be represented by light-years. In actuality, the fastest evolution would not be the enlargement of the cranium. It would be the *development* of the human brain and all of its extraordinary functions, and the appearance of intelligence, and consciousness. All Dawkins wants you to see is the swelling of the cranium, as he says "like a balloon". A balloon is an excellent minds-eye example for Dawkins, as it makes the evolutionary change from ape ancestor to human seem so simple. Evolution must appear simple or it would lose believers. Dawkins also misses that not only did the cranium swell, but the prominent bony brows flattened out, and a prominent vertical forehead had to form to accommodate the much larger frontal lobe of the human brain. The jaws had to flatten out radically as well. Was there a survival advantage to eliminating the large brows and protruding jaws of apes? If so, why did they evolve into existence in the first place? For Dawkins story of evolution from 500 cc to 1400 cc to have occurred, the brain had to expand from about 33 billion brain neurons (brain cells) to 100 billion neurons, an average increase of 335,000 neurons per year if "cranial ballooning" occurred over 200,000 years. The ballooning actually had to take place by generation. If the average generation were 10 years, the increase had to be about 3,350,000 neurons per generation. To add to Dawkins' problem, the brain's glial cells, which are branched cells that support neurons, had to expand from about 30 billion to 80 billion cells, an increase of 250,000 glial cells per year, or 2,500,000 per generation. Can you imagine the increase in blood supply that was required? Hundreds of thousands of capillaries would also have to form every year to feed oxygen to the newly evolved glial cells and neurons. Is it plausible that naturally selected mutations could be responsible for these fantastic increases? Do these increases occur anywhere in nature today?

I have a lot more questions than these for Pope Richard Dawkins. The evolution of Australopithecines to Homo sapiens, if that occurred, had to be much more involved than the cranium enlarging "like a balloon", as he says, "only .01 cc per generation". He is blinded by his own illusions. It's so obvious that he's trying to fool his audience, and he has done a great job of it. He said,

It is absolutely safe to say that if you meet somebody who claims not to believe in evolution, that person is ignorant, stupid or insane, or wicked, but I'd rather not consider (wicked).

I'm sure glad he'd rather not consider wicked. At least I'm not wicked. Maybe.

After Darwin died, paleoanthropologists and their diggers frantically dug in Europe, Africa, and Asia in an effort to find fossils that were mankind's predecessors; fossils that would prove his theory about the origin of man. In a way it was similar to the race to the moon in the 1960's. The hunt for ape-persons and cavemen that preceded humans captured the public's imagination. Evolution supporting scientist and writers were so distraught with the complete lack of any kind of transitional species that showed evidence of an evolutionary lineage from apes to humans that they named the non-found precursor fossils *the missing link*. The public was glued to the search for the "missing link". Whoever turned up the first fossil of a pre-human would certainly be famous, and rich. The frantic search and the incredible desire to be the first finder of a human precursor fossil produced a very strange mix of finds.

For thousands of years mankind has been digging all over the Earth, making homes, foundations, buildings, mines... Along with those digs came a search for fossils of early species. Out of all of these digs, one would think fossils proving ape to human evolution would have shown up in droves. If evolution is valid, there should be millions of pre-human fossils buried all over Africa and Asia. They should have turned up way before Darwin's *Origin*; not as a result of it. They should have been in existence before Darwin was born.

As the 20th century approached, the only early human fossils unearthed were *Cro-Magnon man*, and *Neanderthal man*. Neither could be classified as a missing link, as they were both early humans. Until more impressive pre-human fossils could be turned up, these two guys had to hold down the fort as the only transitional forms that showed evolution from ape to human. If evo-illusionists couldn't continue fooling their audiences when they were devoid of ape to human fossils, evolution was on the verge of collapse. The clock was ticking. Questions would be asked. People might actually become skeptical of evolution, and start thinking on their own. So you can see why the search for the missing link was so frantic. Something needed to be done to keep evolution alive with the general public.

The story of the discovery of human precursors is a fascinating one. Early humans exist in a kind of fog to most people. The fog is created by the complex names given fossil finds, the lineup of different species, the similarity of those species, and the way they're portrayed. *Australopithecus afarensis lead to Homo erectus, which lead to Homo antecessor...* If we were talking about species with names like *elephants, lions, and zebras*, animals with simple single-word names, non-scientists would much better understand human evolution. The complex names alone put human evolution over the audience's heads. It's almost as if human evolution is an immense prank. The evo-illusionists who perform them probably huddle up behind closed doors and have a great laugh at the expense of their gullible audiences that they so easily fool. Most people just figure they will believe whatever evo-illusionists say. I certainly did for many years. It's too much effort to try to figure out which hominid was which, and where they fit on the human tree of life. When I was an evolution believer, any new find, any new hominid fossil that I read about in the newspaper, was monumentally exciting for me. But each one would gradually fade into the fog, and I would soon forget its complex name and where it fit on the tree of human evolution

Forming the illusion that early apes evolved into modern humans requires very definitive steps that have evolved over the last 150 years. Some steps worked, and some didn't. Evo-illusion itself utilized a sort of survival of the fittest scenario. The steps that didn't work well were set aside. The ones that worked survived, and are used today by evo-illusionists all over the world. This series of steps is used to create and promote just about every hominid illusion currently accepted by the audience. Later in this book I will example these steps and show how they were and are followed almost exactly with each hominid fossil find. These are what I call, **Ten Steps to Hominid-ism:**

(1) **The Find:** A few teeth are dug up, or one or a few bones, in Africa or Asia... or Europe; preferably Africa since it has been declared to be the cradle of mankind. The find doesn't have to be much. Converting the bones or teeth into a real-looking hominid will be taken care of in later steps.

(2) **The Hominid Declaration:** The teeth or bones are declared to be hominid, a human precursor, at least hundreds of thousands to millions of years old. If numerous evo-illusionists support the contention and agree that it's a

human precursor, how could anyone in the audience doubt? It matters not that there is no possible way to prove that any fossil find evolved into a human.

(3) **The Tools:** A search is undertaken around the find for stones that might look like tools. If found, they are declared to be tools. The stones are given names, such as *hand axes* and the like, which adds great value to the find, and fame to the finder.

(4) **The Complex Name:** The bones or teeth are given a scientific and very confusing name. This step is key. It fools the audience into thinking there is great value in the find; and that it is such a complicated find that its true nature is way over the heads of the audience. *Australopithecus afarensis*, or *Sahelanthropus tchadensis* are good examples. Think of how confusing these names will be for museum patrons and ordinary citizens. Complex and confusing names cause ordinary people to forfeit their skepticism and thinking to evo-illusionists. They will simply allow evo-illusionists to do the thinking and study for them about hominids like *Sahelanthropus tchadensis*. It's way too complicated of a subject for ordinary people, museum visitors, and the like.

(5) **The Cute Nickname:** The find is given a cute nickname; one the audience will be certain to remember. "Lucy", or "Ardi" are great examples. People will remember these names, and bypass the underlying but dominant complex names.

(6) **The Age Determination:** The teeth or bones are aged using knowingly inaccurate aging techniques. In most cases questionable age determination processes will agree with the finder's estimation of the age of the bones or teeth gleaned from the declared layers or location the find.

(7) **The Addition of More Bones:** Additional bones are added to the find. The skeleton is made as complete as possible. It doesn't matter where in the world the additional bones come from. They could be bones dug up thousands of miles away from the original find. The audience will have no idea the bones are a conglomeration of different finds from different locations. In fact they wouldn't want to know; they would be disappointed.

(8) **The Models and Paintings:** This is possibly the most important step. No matter how partial or minuscule the fossil teeth or bones discovered are, they are next fabricated into full pre-human models and images. Foreheads are added where none existed on the fossils. This step makes hominids look like they had a large frontal lobe, and that they were near human. Craniums are widened so they are outside of the lateral extents of the eye sockets, unlike the fossils. Human-type skin, beards, hair, eyes, and ears are added to make them

look as near-human as possible. Evo-artists are the key step in this illusion. Unsuspecting people who look at the models, drawings, and paintings, are then very easily convinced the fossils are just steps from evolving into humans. Amazingly, few people notice the models and paintings don't follow the shapes of the fossil bones at all. This gives the evo-modelers great latitude in making their models look human-like. Modelers and artists are a huge part of the evo-illusion of man.

(9) **The Timeline:** The fossil hominid is placed on timelines throughout the world. If the finder is famous enough, he can shove other hominids off the timeline, or to minor branches, and move his into its place. It will anger the owner and finder of the moved hominid, but here is where power and true natural selection takes over. The most powerful evo-illusionists get the top spots on the timelines and trees of life. Kind of like survival of the fittest, only with paleoanthropologists.

(10) **The Promotion:** The find is disseminated in every major newspaper and periodical in the world; then in documentaries and textbooks. National Geographic is always helpful here. Ditto PBS and the BBC. They don't question or check the validity at all, since there is no possible way to check. They are an easy sell. Once the big organizations accept the teeth or bones as hominid, the fossil is on its way to fame; and so is the finder. If it gets on the cover of NatGeo, or in a NatGeo or PBS documentary, it has it made.

Now you know the inner workings and secrets of human evo-illusion. Complete plaster and plastic model skulls made from the most partial of skull

Fig. 1-15

bones are on the shelves of natural history museums all over the world. Shelves, tables, drawers, and exhibits full of skulls and other bones are displayed as background on most TV documentaries about human evolution, which gives the illusion that tens of thousands of hominid fossils exist when few actually do. It makes it look like the evidence for human evolution is overwhelming. I certainly fell for the illusion when I was an avid fan of evolution.

I was certain there were innumerable hominid fossil finds, and uncountable skulls. Why, every museum was full of them. I concluded, without skepticism, that all of those skulls on shelves and tables were hominids that evolved into humans; as I was supposed to. It's just a given; a natural unspoken inference. Evolution documentaries display copies and mock-ups of a miniscule few finds, with many ape skulls thrown in. Figure 1-15 is a perfect example. The numerous skulls on the shelves and on the table are *models* of hominid skulls; or they're modern ape skulls. The skulls in the foreground are chimpanzee skulls. The photo is from the NOVA documentary *The Dawn of Humanity*. If they were real hominid fossils, ask yourself why on Earth would these scientists have all of those priceless skulls sitting on the table? One misstep could knock some of these priceless skulls off, causing them to be smashed to smithereens. Look at the skull on the corner of the table. Gad. If it were a hominid skull and worth tens of millions of dollars, it sure wouldn't be sitting there. That's only my guess, but if I had to bet, the skulls are purposefully placed the way they are to support evo-illusion. The big question is, what the heck are these three supposed scientists even doing with these skulls? They're sitting in front of their computers, which looks scientific, and studying the skulls, which looks scientific as well. Unless they're students taking an anatomy test, why are they so fervently studying the skulls? The illusion is that these "scientists" are spending their time advancing science and evolution, and doing real scientific stuff. In reality, this is a posed photo of people putting on airs that they're spending lots of time digging deep into human evolutionary history. It's just part of the illusion.

Evo-illusion's artists and sculptors take over and make the skulls look like they're near human. Certainly no one could challenge the scientific modeling done by evolution's artists and sculptors. Everyone knows how accurate post-mortem artists are in recreating the faces and anatomy of deceased victims for police departments. For sure evolution's sculptors and artists are just as precise when they reconstruct hominids. Aren't they? In the next few chapters I will answer that question. I will review the most important fossils that, according to evolution scientists, prove the evolution of humans from early apes. You will see that each hominid has its own fascinating story.

Chapter 2

"Who Are Those Guys?"- Butch Cassidy

How can the bodies of animals be contrived with so much art, and for what ends were their several parts? Was the Eye contrived without Skill in Optiks, and the Ear without Knowledge of Sounds?... And these things being rightly dispatch'd, does it not appear from Phenomena that there is a Being incorporeal, living, intelligent...?-Sir Isaac Newton

After Darwin died in 1882, paleoanthropologists and their diggers frantically dug in Europe, Africa, and Asia in an effort to find fossils that were mankind's predecessors; fossils that would prove his theory about the origin of man. In a way it was similar to the race to the moon in the 1960's. The hunt for ape-persons and cavemen that preceded humans captured the public's imagination. Evolution-supporting scientist and writers were so distraught with the complete lack of any kind of transitional species that showed evidence of an evolutionary lineage from apes to humans that they named the non-found precursor fossil *the missing link*. The public was glued to the search for the missing link. Whoever turned up the first fossil of a pre-human would certainly become famous; and rich. The frantic search and the incredible desire to be the first finder of a human precursor fossil produced a very strange mix of finds.

For thousands of years mankind has been digging all over the Earth, making homes, foundations, buildings, mines... Along with those digs came a search for fossils of early pre-human species. Out of all of these digs, one would think human fossils exhibiting ape to human evolution would have shown up in droves, either purposefully, or accidentally. They didn't.

As the 20th century approached, the only early human fossils that had been discovered were named *Cro-Magnon*, and *Neanderthal*. Until more impressive pre-human fossils could be turned up, these two guys had to hold down the fort as the only transitional forms that showed evolution from ape to human. If evo-illusionists couldn't continue fooling their audiences when they were devoid of human precursor fossils, evolution was on the verge of collapse. The clock was ticking. Questions would be asked. People would actually start doubting and thinking on their own. So you can see why the search for the missing link was so frantic. Something needed to be done to placate the public.

Fig. 2-1

Neatherthal man: From the time Darwin published his book in 1859, *On the Origin of Species*, he had no early pre-human species that he could cite that evolved into humans. Fossils from fish and plants *hundreds of millions of years old* were plentiful. But the fossils that demonstrated the evolution of humans from ape precursors that were only *a few million years old* didn't exist at all. Darwin's only hope was the unearthing of *Neatherthal Man* (Figure 2-1) in the Neander Valley of Germany. Neatherthal Man was discovered in 1853, three years before Darwin published *On The Origin of Species*. Neatherthal Man has been, evolutionarily, a gray area human precursor since his uncovering. Soon after the discovery of Neatherthal, a University of Berlin professor and leading pathologist of his day, Dr. Rudolf Virchow, determined that that first Neatherthal skull was just an unfortunate Homo sapiens who had suffered childhood rickets and adult arthritis. His skull had also been "modified" by several severe blows. Was the poor bloke that owned that skull murdered?

Neatherthals walked upright in the same manner as modern humans. There's a good deal of evidence to show that Neatherthals cared for the sick and old in their communities. Diggers found other Neatherthal fossils that showed potentially life-threatening injuries, which had healed, indicating that other family members nursed the Neatherthal, who suffered the injuries, to health. Ancient musical instruments were found in Neatherthal digs, which indicates Neatherthals played music. Their brains were as large as our brains;

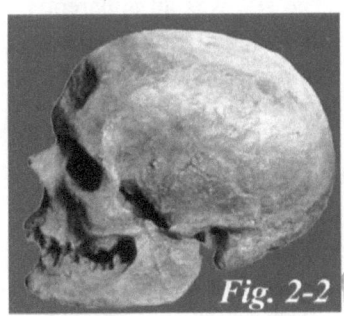

Fig. 2-2

they had substantial foreheads, and large ovoid craniums. It was thought that all Neatherthals lived in caves, which fit with the notion that the first humans were cavemen. In reality, many of them lived in huts, similar to American Indian teepees, using branches and bones covered with animal skins. For all intents and purposes, Neanderthals were very much human, and lived during the same time as modern humans, which eliminates them from being a possible human precursor.

Neanderthal's could procreate with humans, which means they were the same species. Boy, wouldn't that be a sexy evening? In fact they may have not gone extinct at all. They may have simply blended in with humans. Neanderthals were either disfigured humans, or just another race of humans. According to paleoanthropologists, Neanderthal man lived from about 200,000 years ago until about 40,000 years ago in Europe. The fact that Neanderthals were discovered in Germany didn't help at all, since early man supposedly evolved in Africa. Humans didn't leave Africa until about 60,000 years ago. So how did Neanderthal man wind up in Europe 140,000 years before Homo sapiens left Africa? Evolution's own "facts" make the entire human family tree as proposed by evo-illusionists a complete mess. [1-4]

Cro Magnon: The next big fossil find that was promising as evidence for human evolution was *Cro-Magnon man*. (Figure 2-2) I remember when I was in university biology and anthropology classes; I really liked Mr. Cro-Magnon. He seemed acceptably human. He made me feel more comfortable with the evolution of man. The other hominid fossils were a bit too ape-like. I was a firm evolution believer, but deep down I did have trouble with species that looked like apes, yet they were credited with being human precursors. So Cro-Magnon was my favorite early man. Cro-Magnon was artistically placed in a lineage of human precursors, species that evolved step by step into Homo sapiens, because he was sorely needed. A road construction crew found Cro-Magnon in 1868 in France. They had unearthed a rock shelter built into a limestone cliff. In the back of the shelter were the remains of four adult skeletons, one infant, and some fragmentary bones. The site was an apparent gravesite. There were ornaments buried with the fossils including what looked like pendants and necklaces. Cro-Magnon had all of the characteristics of Homo sapiens. He was initially considered to be in the hominid lineage to human beings, but his age, unfortunately, made that impossible. Modern dating techniques say that Cro-Magnon lived between 43,000 to 32,000 years ago. They were early "Frenchmen", and not early hominids at all. If his age is correct, Cro-Magnon fits anthropology's timeline that says humans migrated out of Africa 60,000 years ago. This means Cro-Magnon had to be human, since, according to evo-illusionists, fully evolved humans first appeared

200,000 years ago. Cro-Magnon had to be removed as a human ancestor on the tree of human lineage that was accepted as valid when I was in school in the 1960's. Cro-Magnon *was* human, not an evolutionary human ancestor. One paper on Cro-Magnon stated, *Further research, over the past 20 years or so, showed that the physical dimensions of Cro-Magnon aren't sufficiently different enough from modern humans to warrant a separate designation.*[5]

Why did it take evolution scientists 130 years to Figure that out? Couldn't they measure bones 130 years ago? Anyone could use the information I gave in Chapter 1 about ape and human skulls to determine Cro-Magnon was human; anyone except a needy evo-illusionist. Cro-Magnon had flat brows, a large ovoid cranium, a large forehead, and flat jaws, which makes him very human. Why couldn't scientists define Cro-Magnon for all of that time? Was Cro-Magnon considered to be an ancestor of humans because there weren't any human ancestors in the fossil record? That's my bet. Evo-illusionists desperately needed Cro-Magnon to support their illusions. So he was bent and exacerbated. At any time during those years any scientist could have taken the "measurements" that determined Cro-Magnon was, indeed, human. Cro-Magnons are now designated *Anatomically Modern Human* (AMH) or *Early Modern Human* (EMH). In any case, they were humans. The name Cro-Magnon was discarded.

For thirty years after Darwin published Origins, there were no fossils that could be cited that proved mankind came from earlier simpler forms. In early1882, Darwin was diagnosed with severe heart disease. He died of heart failure at his Down House on April 19, 1882. The poor bloke never did see any of the fossil digs that came after his death that supposedly supported his contention that man evolved from an ancient primate species. At the time of his death, all he knew were Neanderthals, and Cro-Magnons, neither of which did him a lick of good in proving human evolution. He had no idea about the effort his fellow believers and supporters would exert to promote his theory. I don't know if he'd be proud. My bet is he'd be embarrassed.

Frantic searches and digs ensued to find man's missing link. Paleoanthropologists had to go to the immense continents of Africa, Europe, and Asia and dig in the dirt, gravel, and layers of shale in hopes that they might come up with a human precursor. Keep in mind that at the time of Darwin's death, no human *precursors* had been found; well, other than Neanderthal and Cro-Magnon. Where do you start digging, on a continent as immense as Africa? Just think how tiny your shovel would look; or even your steam shovel.

Digging up a missing link, or lots of missing links, would be like looking for a needle in thousands of haystacks. A very difficult task, particularly if the searchers weren't even sure the needle existed. Just as it seemed all was lost, just in the nick of time, an immense find was made.[6-9]

Fig. 2-3

Piltdown Man: Paleontologists looking for human precursors dug for decades, and never came up with much more than old dog bones. The longer they dug without results, the more they spent their grant money, the worse their failure looked. So they had great proclivity to come up with something… anything! Otherwise, they'd lose their funding, and wind up broke, with nothing to show for all of their work but the egg on their faces.

One of the world's greatest pranksters must have noticed the blind insatiable hunger paleoanthropologists had for being the first to uncover a true missing link. In 1908 he took bones he collected from different sources and assembled them so they resembled an ape and human mix; a human ancestor. He did such a horrible job of making his fossil ape-man, anyone with half of a brain could tell it was a fake. He surreptitiously buried the fake fossil in a gravel pit near Sussex, UK. Then, in front of witnesses, he dug it up. (Figure 2-3) He feigned great surprise and excitement. His witnesses may have been a team of pranksters who were in on the prank to help support his find.

The New York Times.

SUNDAY, DECEMBER 22, 1912.

C

DARWIN THEORY IS PROVED TRUE

English Scientists Say the Skull Found in Sussex Establishes Human Descent from Apes.

THOUGHT TO BE A WOMAN'S

Bones Illustrate a Stage of Evolution Which Has Only Been Imagined Before.

CREATURE COULD NOT TALK

Probably Lived at a Time When Other Species of Humans Had Developed Further Elsewhere.

Special Cable to THE NEW YORK TIMES.

LONDON, Dec. 21.—A race of apelike and speechless men, inhabiting England hundreds of thousands of

race, but the bosses of bone in the Sussex man are smaller than in the higher race.

But it is not, Dr. Woodward declares, till we come to an examination of the lower jaw that the full significance of the discovery becomes apparent, for while the brain case is emphatically human the jaw is as emphatically apelike. Found by itself it might and would be regarded as that of an ape with many human features in its general conformation.

This remarkable fragment agrees with the celebrated jaw found about five years ago on Mauer Heidelberg and known as the Heidelberg jaw, but it presents an apelike feature which the less ancient Heidelberg jaw does not. The most striking point of both is the extraordinarily receding chin, the jaw sloping backward sharply from the base of the teeth, which had a decided forward thrust.

In the living races of mankind the chin is always more or less conspicuous, the lower border of the jaw standing well in advance of the teeth, which are mounted vertically along its rim.

Other apelike features of the jaw are the absence of the muscular ridge along its inner surface, known as the mylohyoid ridge, affording attachments for the muscles of swallowing and speech. This ridge is always present in the human jaw.

Finally in apes two branches of the lower jaw, where they meet in front

Fig. 2-4

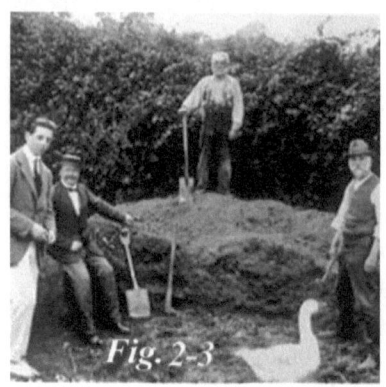

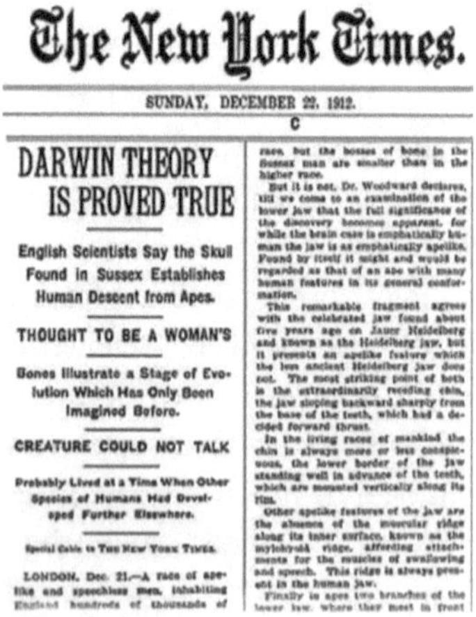

On a cold foggy February morning in 1912 (I know because all February mornings are cold and foggy in the UK), Arthur Smith Woodward, head of the geology department at the Natural History Museum in London, received a letter from Charles Dawson, the probable head Piltdown prankster. Dawson was a lawyer and amateur fossil hunter. The letter dropped a bombshell. He wrote that he had stumbled on a very old layer of gravel, near a village called Piltdown, where he had found iron-stained flints and "a portion of a human skull". On request from Woodward, Dawson brought the fossil bones to London where they were perused by numerous paleoanthropologists. Fragments of Piltdown's fossil skull and jawbone were displayed at a meeting of the London Geological Society. They were unveiled to the world with great fanfare. Piltdown's fossil fragments were quickly claimed to be "the earliest Englishman - Piltdown Man!". Darwin's theory on the evolution of man was saved! Piltdown man was officially given the complex name *Eoanthropus dawsonii* after its erstwhile finder. As I said earlier, giving a pre-human fossil a complex name like this renders it sacred. It's a very important step in the illusion. Once it has a complex name, no one can challenge its veracity. The name alone gives it authenticity. On December 18, 1912, newspapers throughout the world ran sensational headlines, like the one in the New York Times. (Figure 2-4) Once Piltdown Man, or *Piltdown Prank-man,* as I have renamed him, was launched, there was no turning back.

Was Dawson himself the prankmeister? Since the prank has been uncovered, over 30 people have been accused of being the prankster. There is no doubt, though, that Dawson was either the sole conspirator, or he worked with others to concoct the prank. In the 1970's, sixty years after his death, his initials were found on an old canvas traveling trunk, hidden in a museum loft, that contained mammal teeth and bones stained and carved in the manner of the Piltdown fossil. Actually, when it comes to suspects, the Piltdown Prank-man makes *Midsummer*

Fig. 2-5

Murders look like chicken feed. There was probably a whole team of pranksters.

It's certain that whoever perpetrated this prank had no idea that it would go as far as it did. Evo-illusionists like to call this incident the greatest scientific hoax in history. But the perpetrator did such a horrendous and obvious job of faking the fossil, it surely couldn't have been anything more than a prank. It didn't qualify as a hoax. Take a look at the photo (Figure 2-5) of gullible scientists checking out the sacred site where Piltdown Prank-man was concocted; that is, *earthed*, then *unearthed*. The perpetrators probably thought the paleontologists who studied the gag fossils would recognize they were fake, and the prank would be quickly dismissed. Everyone would get a good laugh, and that would be that. "Hey boys, let's open some ale!" For his next prank maybe he had a Whoopee cushion to put under one of the paleontologist's seat cushions; hey, and maybe disappearing ink! When the fossil jokes were unearthed and taken seriously, the pranksters must have been in disbelief. They also must have had a good chuckle, and been shocked that anyone would take the joke so seriously. So they just let it ride. In fact, between 1908 and 1915 the Piltdown site was continually salted with more "fossils". Paleoanthropologists kept digging them up and being fooled by the obvious fakes; and the prank kept expanding. Piltdown Prank-man was hailed all over the world as *the missing link*. Piltdown Prank-man was way past the point of no return. So the pranksters just let it ride, and kept quiet. Anyone would have done the same. Right? I sure would have. How could they spoil everyone's excitement? Then when renowned paleoanthropologist Sir Arthur Keith spent five years of his life studying the prank fossil, the pranksters were really in deep. They just couldn't admit to this *prank gone wild*. So they sat back and relaxed, and enjoyed the ride. Whoever the prankmeisters were, the odds are they produced many more pranks. Someone who would produce such an incredible prank usually starts with smaller ones. When the smaller ones work, and then when the large one is a raging success, it usually

Fig.2-6

spurs the prankmeisters on to more pranking. It's usually the tip of the iceberg. In fact that appears to be the case.

Piltdown Prank-man quickly found its way into textbooks (Figure 2-6), encyclopedias, and university lectures. Evolution scientists throughout the world treated Piltdown Prank-man with great reverence. Did the illusion really fool them too? Or did they use Piltdown Prank-man to expand their own illusions. Evolution's artists and sculptors, of course, had a field day. They took the prank and evolved it into a fully formed early hominid. Artistic charts

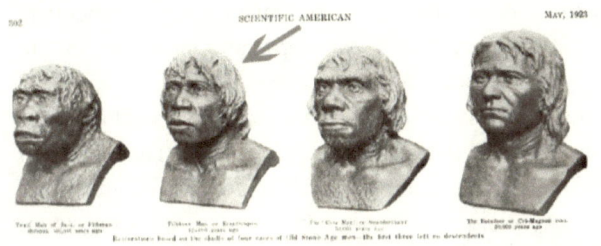

Fig. 2-7 Extinct Races of Ape-Like Man
Tracing the Evolution of Man and the Apes from a Common Ancestor

showing the evolution of man now included Neanderthal, Cro-Magnon, and Piltdown Prank-man. Figure 2-7 is from the May 1923 issue of *Scientific American*. It wasn't so scientific in this case, now was it? Scientific American displayed Piltdown Prank-man (arrow) in the series that represented ape-to-man evolution. The two on the right are Neanderthal man and Cro-Magnon. Scientific American unashamedly and proudly included itself in the large group of gullible pushovers fooled by the Piltdown prankmeister. The evolution of man from apes was proved! Piltdown was accepted!

"Piltdown man was a really big deal in 1912, because it was a time when very little was known of human fossil remains", said historian Richard Milner. It was perceived to be the missing link, the fossil that connected humans with apes. Notably, Piltdown Prank-man was even more accepted and famous than the celebrated Neanderthal and Cro-Magnon men.

Astoundingly, no scientific tests were run on the skull and jaws. If scientists had done testing, they would have noticed the chemical staining and filed-down teeth very quickly. The staining was pathetic; it looked like furniture stain, which it probably was. Testing would have shown that Piltdown Prank-man was not a genuine artifact. It would have collapsed quickly. Everyone would have had a good laugh, and that would have been that. Evolution scientists accepted it because they were in dire need. Without Piltdown Prank-man, they were back to square one as far as evidence for human evolution went. English researchers, in particular, were overly willing to accept Piltdown Prank-man finds. Not only was scientific illusion involved in promoting the prank, but English nationalism was as well. Paleontology in Britain was going through very lean times. British paleoanthropologists desperately wanted to believe that fossil gold had been unearthed. Cro-Magnon found in France and Neanderthal found in Germany had left England out in the cold. If France and Germany were such important stepping stones for human evolution, England certainly should have been one too. Up to this point in time,

Britain had nothing. One French paleontologist had even disdained his English counterparts as *merechasseurs de cailloux* – pebble hunters. English scientists wanted to show that their country had been an important incubator in the formation of Homo sapiens, just as Germany and France were.

Mr. Piltdown had a skull that was obviously human, and a jawbone that looked like it came from an ape. But it had human-like teeth. What a mix! In fact, Piltdown Prank-man consisted of a human skull mated with a mandible

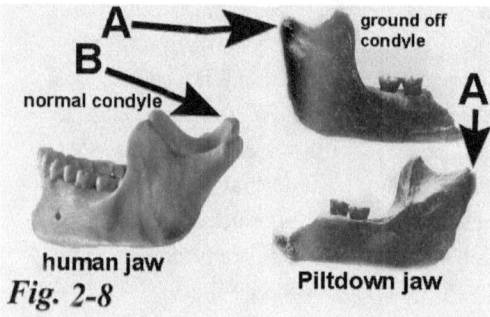

Fig. 2-8

(lower jaw) of an orangutan. The pranksters heavily altered the teeth and bones so they would match up as closely as possible. In reality, it wasn't possible to match an orangutan jaw with a human skull, so parts of the mandible that might reveal the mismatch were removed. Note in Figure 2-8, the condyles, the *ball* of the ball and socket joints, (A) were ground off. Compare them with the condyles (balls) of a normal human condyle (B). Didn't any paleontologist that handled Piltdown Prank-man wonder why *both* condyles were equally absent? Wasn't the reason obvious? The jaw was colored with stain to match the skull. The teeth of the mandible were ground to match those of the maxilla (upper jaw). But even that would have been a gross failure, because the teeth of an orangutan couldn't be ground to match the teeth of a human without being completely obvious. You are now an expert as to why that's the case. The human flared dental arch could not fit the "U" shaped arch of an ape. The canine teeth (eyeteeth) were reshaped to make them look properly worn and human. Any dentist with moderate training could have uncovered the hoax, even in 1908. Wasn't the Piltdown fossil shown to a dentist? Of course not. Why rain on the parade and ruin a really great evo-illusion? Piltdown Prank-man was a major prop that could support the illusion of apes evolving into humans. It was a matter of need, not good science.

The skull of Piltdown Prank-man was unusually thick; a condition that is rare in the general population but is common among the Ona Indian tribe in Patagonia at the southern tip of South America. How would one go about getting a skull from Indians who lived at the southern tip of South America, in 1912? I can't imagine. The other bones of Piltdown Prank-man were gathered from a variety of sources. This is an astounding thought, since travel at the time of the discovery of Piltdown Prank-man was difficult. Ships, trains, and horse-drawn carriages were the common mode of travel. Just think how much trouble the prankmeisters went to in order to concoct this prank and fool so many people.

Piltdown Prank-man had a lifespan of over forty-five years; certainly way beyond the prankmeister's wildest expectations. Interestingly, the March 1950 Nature magazine had an article titled *New Evidence on the Antiquity of Piltdown Man*. The author, Sir Kenneth Oakley, who was one of the three paleoanthropologists who exposed the prank only three years later, wrote:

The results of the fluorine test have considerably increased the probability that the [Piltdown] mandible and cranium represent the same creature. The relatively late date indicated by the summary of evidence suggests moreover that Piltdown man, far from being an early primitive type, may have been a late specialized hominid which evolved in comparative isolation. In this case the peculiarities of the mandible and the excessive thickness of the cranium might well be interpreted as secondary or gerontic developments.

So only three years before Oakley worked to expose Piltdown, he himself was touting it as a valid hominid. Why did the orangutan jaw, that didn't fit the skull at all, fool him so? In July 1953 an international congress of paleontologists convened in London. It was at that congress that the possibility of fraud dawned on paleoanthropologist Joseph Weiner. Once the possibility was raised it was easy to establish that Piltdown Prank-man was, indeed, a prank. Ronald Millar wrote in his book, *The Piltdown Men*:

*The original Piltdown teeth were produced and examined by the three scientists. **The evidence of fake could be seen immediately.*** (Me: Immediately? So why didn't the first scientist that saw the fossil decades earlier see it was fake immediately? And the dozens of scientists who studied it after those first three?) *The first and second molars were worn to the same degree; the inner margins of the lower teeth were more worn than the outer -- the 'wear' was the wrong way round; the edges of the teeth were sharp and unbevelled; the exposed areas of dentine were free of shallow cavities and flush with the surrounding enamel; the biting surface of the two molars did not form a uniform surface, the planes were out of alignment.*[14]

The molar surfaces were examined under just an ordinary magnifying glass. Scratches on the surface had a crisscross pattern, suggesting an abrasive was used to do the reshaping. True wear lines from chewing and tooth-to-tooth abrasion would have left parallel scratches, not crisscross patterns. Why did it take over 40 years to examine the teeth with a magnifying glass? Le Gros Clark, a paleontologist at the meeting who studied the Piltdown fossil said:

The evidences of artificial abrasion **immediately sprang to the eye.** *Indeed so obvious did they* [the scratches] *seem it may well be asked --* **how was it that they had escaped notice before?** *They had never been looked for...nobody previously had examined the Piltdown jaw with the idea of a possible forgery in mind, a deliberate fabrication.*[14]

There were so many evo-illusionists who supported this prank that it didn't matter how amateurish the modifications were. Evo-illusionists are always on the hunt for more illusions, as are all illusionists and magicians. A missing link was sorely needed; Piltdown Prank-man was it. It didn't matter if Piltdown was valid or not. Who was going to kill Piltdown Prank-man and the illusion that went with it? Whoever did would certainly lose their grants and be scorned. Piltdown was just about the only prop available that could support the illusion of the evolution of mankind for many years. By 1953, diggers had come up with other hominid fossils that were utilized as props for evo-illusion, even though they were also questionable. So in 1953 Piltdown Prank-man could be exposed without causing the complete collapse of the illusion of human evolution, and without causing scientists that did the exposing to be shunned.

temperature"—68 degrees above

(Continued On Page 15A)

Piltdown Man Named Hoax, Jolts Science

London, Nov. 21 (INS)—Three British scientists branded the famous "Piltdown 'man" a deliberate fake today and set off a controversy that may engage the scientific world for years to come.

Lengthy Investigation

On the basis of lengthy investigations the British Museum's bulletin called it a "hoax without parallel in the history of paleontological discovery."

The skull, jaw, and teeth were found in 1912 when a gravel pit was dug at Piltdown in Sussex County.

Despite the nagging doubts of some authorities over the fragments' authenticity, other scholars based long studies of what man was like 50,000 years ago on the Piltdown findings.

The report released today by Drs. J. S. Weiner, K. P. Oakley of the British Museum and Prof. W. E. Le Gros Clark of Oxford University says there is "no doubt that whereas the Piltdown cranium may well be of the Upper Pleistocene" period, the jawbone and

(Continued On Page 12A)

Fig. 2-9

The news of the discovery that Piltdown Prank-man was a hoax spread like wildfire. It was truly huge news. *Piltdown Man Named Hoax Jolts Science,* (Figure 2-9) announced the New York Times on November 21, 1953.

Part of the skull of the Piltdown man, one of the most famous fossil skulls in the world, has been declared a hoax by authorities at the British Natural History Museum, the article said. The London Star headlines shouted, *The Biggest Scientific Hoax of the Century.* Actually the London Star was wrong.

The biggest scientific hoax of the century is human evolution itself. Piltdown Prank-man was a smaller hoax within a massive hoax. Several highly respected and serious scientists were made to look like complete buffoons. Their reputations were forever tarnished. Can you imagine the embarrassment of being known for spending years being fooled by such an obvious prank? The New York Times in 1953 further reported,

Sir Arthur Keith, famous British paleontologist, spent more than five years piecing together the fragments of what he called a 'remarkable' discovery. He said the braincase was "primitive in some respects but in all its characteristics distinctly human".

How foolish he must have felt, having worked five years on an obvious prank without figuring out that it *was* a prank. I wonder what he said to his wife at the dinner table the night the headlines came out. "Honey, I have some news....uh. Ya know that fossil I've been working on for the last five years... well...uh, could you make me a peanut butter and jelly sandwich for lunch tomorrow?" To further embarrass Sir Arthur, I thought it would be nice to include his picture. (Figure 2-10)

Fig. 2-10

It turns out that there never were any significant fossils of any kind at the Piltdown quarry, which makes one wonder why there were any digs there in the first place. Evolution scientists and their audiences hate even the mention of Piltdown Prank-man. Their reactions are always in the, "Oh, no, not that old canard again" category. They try to brush it off as only an asterisk in the otherwise wonderful history of the science of human evolution. If you bring the subject up, you are trite, corny, and out of touch. Evo-illusionists try to turn Piltdown Prank-man into a positive for evolution. They tout that evolution, a "real science", corrected a major error, as good science should; and that this is further proof that evolution is truly real science. They unashamedly ignore that the entire world of scientific evolution swallowed Piltdown Prank-man without the slightest bit of skepticism; and they did so for almost a half of a century. Good science is composed of a very strong dose of skepticism, and evolution had absolutely none in the case of Piltdown Prank-man. In fact, all evolution evidence is devoid of skeptical perusal by evolution scientists. They also proudly brag that evolution scientists uncovered the prank, when in truth only three did. Thousands made fools of themselves supporting Piltdown Prank-man for decades. They would probably still be fooled today if weren't for Joseph Weiner and friends. Good science demands skepticism. If evolution scientists had a skeptical eye, Piltdown Prank-man would have crashed and burned quickly. It would have found its place in the dust heap of historical scientific *faux pas* and failures.

Tests were run in 2012 to determine the identity of the prankmeisters. What could possibly be a bigger waste of money and time? All anyone needs to know is that a horribly constructed prank fossil was concocted that fooled thousands of evolution scientists throughout the world for nearly half of a century. What difference does it make who the perpetrators were? They're all dead. No living person knew any of them, or could care less who they were.

Justin Dix, the Southampton University geochemist who carried out much of the chemical analysis said, "The trouble is that after 100 years we still do not know the identities or motives of those responsible. It's time we did." So probably hundreds of thousands of pounds were invested in this sham investigation; another sub-illusion in a massive illusion. The sub-illusion is: *It's important to find out who did the Piltdown pranking.* Why not just say Dawson did it and leave it at that? I give him tons of credit for being a pretty cool dude; for fooling so many evolution scientists who themselves have dedicated their lives to fooling so many people.

Fig. 2-11

As I said, the first finder of an acceptable missing link fossil would become world-famous and rich. Dawson (Figure 2-11) was certainly both after the Piltdown Prank-man was in gear and rolling along. He was celebrated as one of the world's greatest paleoanthropologists. He would have been knighted, had he not died of septicemia in 1916. He never knew how his prank turned out. Incredibly, both Keith and Woodward were knighted because of this amazing ruse. It turned out Dawson had many secrets. He made at least 38 hoaxes or dodgy artifacts before Piltdown Prank-man that were discovered after his death. He forged axes, statuettes, ancient hammers, Roman tiles and a host of other relics. He was so prolific, believed, and respected that, even though he wasn't knighted, he was awarded fellowships of both the Geological Society and the Society of Antiquaries. So you see Piltdown was not just a one-shot prank. It was the tip of the iceberg. Dawson apparently was a mighty prankster, a fun loving guy, and even a pool player. Maybe Dawson was buried with his disappearing ink, his Whoopee cushion and his snappy gum... and a big smile on his face. The Piltdown Prank-man is certainly one of the best examples of how hungry evo-illusionists are for evidence. It really doesn't matter how fake the evidence may be. Remember, this illusion was created and supported by the world's evo-illusionists. Which proves that as long as most evo-illusionists agree any evidence is valid, no matter how absurd it may be, it stands. [10-19]

Nebraska man, AKA Nebraska Tooth-man

The second most massive illusion within the field of human evo-illusion was *Nebraska Man,* or what I call *Nebraska Tooth-man.* Why? Because he was comprised of a single tooth! Yes folks, an entire human precursor made to appear out of just one tooth. What an illusion! I love this guy. He's just the epitome of evo-illusion. Evo-illusionists do their thing, knowing that their audiences will believe whatever they say, no matter how absurd. And what could possibly be more absurd than making an entire pre-human out of a single tooth?

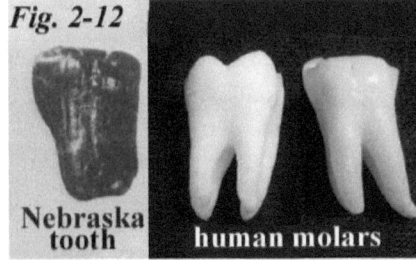

Fig. 2-12

Nebraska tooth   human molars

In 1917, a rancher and geologist, Henry Cook, found a human-like tooth (Figure 2-12) in ancient sediments in Snake Creek, Nebraska. Fully five years after the big find, in 1922, Cook sent it to Dr. Henry F. Osborn, head paleoanthropologist at the American Museum of Natural History, in New York. It's so strange that five years elapsed between this earthshaking find and the sending of the tooth to Osborn. I wonder, if I found a tooth in my backyard, and sent it to the American Museum of Natural History, could they make it into a whole human precursor? Is that offer still available? I'd like to sign up. Actually, if I did find a human-like tooth in my backyard whilst digging up plants, at best it would wind up in my junk drawer. I'd Figure my gardener or a workman dropped it; or maybe an animal. I sure wouldn't Figure it was from a million-year-old hominid. So why on Earth did Mr. Cook send the tooth to the

American Museum of Natural History fully five years after he found it? What a puzzle that is. There must be an entire side story here. Further, anyone could tell the tooth wasn't from a human or even a human precursor. Check out the human molars to the right of Nebraska tooth. Would you have any trouble telling that the tooth touted as an ape-man's tooth wasn't that at all?

Well, Nebraska Tooth-man was bestowed with a complex name, and this one is a doozie. He became *Hesperopithecus haroldcookii*. Evo-illusionists the world over accepted the declaration that he's the first anthropoid ape from North America. Can you imagine? How the heck evolution scientists thought Nebraska Tooth-man got from Africa to Nebraska is anyone's guess. Maybe he took a 747? Osborn received the tooth on March 14, 1922. He wrote to Cook,

I sat down with the tooth and I said to myself: 'It looks one hundred percent anthropoid. (human precursor). [21]

Wow! Well, 100 percent anthropoid? That sure locks things up. Of course, a full reconstruction was commissioned based on nothing more than the tooth found by

THE EARLIEST MAN TRACKED BY A TOOTH: AN "ASTOUNDING DISCOVERY" OF HUMAN REMAINS IN PLIOCENE STRATA.

Fig. 2-13

Cook, and the desire to produce a new illusion of human evolution. Very soon, illustrations of an ape-like subhuman appeared in books and newspapers worldwide. In a 1922 issue of *Illustrated London News*, an article was published featuring a drawing of Nebraska Tooth-man. A tooth found in Nebraska was transformed into an entire human precursor using the magic of evolution art. (Figure 2-13) The artist was Amedee Forestier. The article stated, "the reconstruction is merely the expression of an artist's brilliant imaginative genius." Imaginative? Yes. Genius? No. Huge illusion? Yes. Nebraska Tooth-man became the *Ape Man of the Western World*. The news was colossal; it went worldwide. It made the front page of the New York

Times. (Figure 2-14) *Science* magazine was euphoric because of the find. In an article from May 5, 1922, it stated:

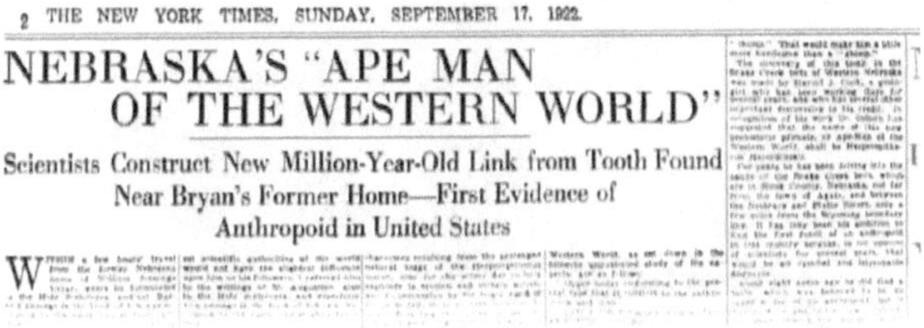

Fig. 2-14

THE NEW YORK TIMES, SUNDAY, SEPTEMBER 17, 1922.

NEBRASKA'S "APE MAN
OF THE WESTERN WORLD"

Scientists Construct New Million-Year-Old Link from Tooth Found
Near Bryan's Former Home—First Evidence of
Anthropoid in United States

It is hard to believe that a single small water-worn tooth, 10.5 mm by 11 mm in crown diameter, can signalize the arrival of the anthropoid Primates in North America in Pliocene time. We have been eagerly anticipating some discovery of this kind, but were not prepared for such convincing evidence of the close faunal relationship between eastern Asia and Western North America as is revealed by this diminutive specimen.[23]

Convincing evidence? A single tooth? Osborn, the museum curator, didn't attempt to discard the meager evidence of Hesperopithecus haroldcookii in his junk drawer, or even in any drawer at the American Museum, as he should have, and as it turned out, wished he would have. For some crazy reason, Osborn had plaster casts made of the tooth. He sent them to twenty-six natural history museums in Europe and the United States. Twenty-six museum curators got a box with a plaster tooth in it. The plaster tooth undoubtedly had a letter along with it telling the curators that this plaster tooth was Nebraska Tooth-man, the *Ape Man of the Western World*. How could any intelligent human who looked at the plaster tooth and who read Osborn's conclusion not think he was crazy? But, serious discussion actually did ensue about whether the tooth really is the *Ape Man of the Western World*. Renowned British anatomist Grafton Elliot Smith jumped into the fray and acknowledged H. haroldcookii as the third known genus of extinct hominids, along with Piltdown man and Pithecanthropus. No one's going to fool Grafton Elliot, that's for sure. What a scientific collection! After viewing one of the plaster teeth, brilliant British paleontologist Sir Arthur Smith Woodward, of Piltdown fame, was skeptical; but not very. Instead of saying, "Are you joking?", he wanted more evidence. I can see why. He said, "The occurrence of a man-like ape among fossils in North America seems so unlikely that good evidence is needed to make it credible." He was fooled by the Piltdown skull, *and* Nebraska tooth; but not entirely.

45

In 1927 further excavations at Snake Creek unearthed more of the Nebraska Tooth-man fossil. They revealed the true identity of its source. Hesperopithecus haroldcookii wasn't the first *Ape Man of the Western World* after all. Hesperopithecus haroldcookii was the tooth of a *peccary*, (Figure 2-15) a South American wild pig. Yes folks, a pig's tooth found in the dirt in

Fig. 2-15

Nebraska was quickly evolved into the *Ape Man of the Western World* by evolution's artists. But then, like in a bad dream for all diligent evo-illusionists of the day, the *Ape Man of the Western World* was quickly evolved into what it really was all along; a South American pig. A major evo-illusion was destroyed with the turn of a few shovels of dirt. I wonder if the rest of the peccary's skeleton hadn't been found, would evo-illusionists still tout Nebraska Tooth-man as a human precursor? Would children all over the world be taking tests and answering questions about the first *Ape Man of the Western World*? I have no doubt they would. Well, there you have it. Finally, after decades of waiting since Darwin's book on human evolution, real visible observable evo-illusion in action. This is a great example of natural selection as well, as the digger had to select where he should stick that shovel in the dirt.

If thousands of people all found a strange tooth in their yard, I can't imagine any of them would send it to a museum of natural history. So Nebraska Tooth-man wouldn't even qualify as a prank. It would fail as a prank, but it sure didn't fail as a much-needed serious hominid illusion for evo-illusionists. Nebraska Tooth-man did its job. As a stopgap, it kept the audience fooled for five years while other digs were being made and new illusions of human evolution were being conjured. For five years that tooth was the first pre-human in North America. The *Ape Man of the Western World*! Evo-illusionists make great fun of the Genesis version of the origins of man, wherein Adam's rib was used to make Eve. Evolution scientists of the 1920's trumped that in spades. They used a pig's tooth to conjure an entire population of male *and* female ape-persons. Sadly Nebraska Tooth-man had a very short life; only five years, compared to Piltdown's 45 years. Can you imagine how embarrassed Osborn must have felt, having worked five years on an obvious pig's tooth without figuring out that it *was* a pig's tooth? I wonder what he told his wife at the dinner table the night after the collapse of *The Ape Man of the Western World*. "Honey, I have some news....uh. Ya know that tooth I've been working on every day for the last five years… well… uh, could you make me a ham sandwich for lunch tomorrow?" To further embarrass Osborn, I thought it would be nice to include his picture. (Figure 2-16)

Fig. 2-16

Nebraska Tooth-man played a role in the world-famous Scopes Trial that took place in 1925 that determined if a teacher could or did teach evolution in the science classroom. In 1922, Osborn was speaker before the National Academy of Science. Osborn said the discovery of Nebraska Tooth-man had come soon after he had:

advised William Jennings Bryan to consult a certain passage in the book of Job, 'Speak to the earth and it shall teach thee," and it is a remarkable coincidence that the first earth to speak on this subject is the sandy earth of the Middle Pliocene Snake Creek deposits of western Nebraska.

He sarcastically suggested that the animal should have been christened *Bryopithecus* (after William Jennings **Bryan**), "the most distinguished primate which the State of Nebraska has thus far produced." He was very sure of Mr. Nebraska Tooth-man. Just before the Scopes trial began, in which Osborn was to be a witness for Darwin and Scopes, Osborn went dead quiet. Further digs at Snake Creek just before the trial turned up the rest of the pig bones, and that was that for Nebraska Tooth-man. Osborn didn't testify at the trial. If he had gone on the witness stand, William Jennings Bryan would have made mincemeat of him.

Were *The Ten Steps to Hominid-ism* followed when the Piltdown Prank-man-man and Nebraska Tooth-man-man illusions were introduced as human precursors? Of course they were. Both were early faked attempts at proving Darwin's theory on the evolution of man, and the steps were almost perfectly followed.

(1) **Find:** Faked skull at Piltdown, UK, and a pig tooth found in Nebraska.

(2) **Declare:** Both were declared to be human precursors, and the declarations were supported by fellow evo-illusionists in droves.

(3) **Tools:** No tools were declared, so this step was skipped. The tools came with later finds.

(4) **Complex Name:** The Piltdown find was christened *Eoanthropus dawsonii.* The Nebraska find was christened *Hesperopithecus haroldcookii.*

---

(5) **Nickname:** These two were dubbed *Piltdown man* and *Nebraska man*. Later "finds" were given cuter nicknames as the technique evolved.

(6) **Age Determination:** Phony estimates were made that fit the story of human evolution.

(7) **Add to the Find:** Additional bones were found at the Piltdown site, placed there by the pranksters. Nebraska Tooth-man-man pretty much remained a single tooth.

(8) **Artwork and Modeling:** Both Nebraska Tooth-man and Piltdown Prank-man had extensive artistic recreations made. Piltdown clay models made appearances in anthropology textbooks. A great example of evo-artwork is Figure 2-17 showing Piltdown Prank-man-man on the hunt.

(9) **Place on a Timeline:** Both were placed on timelines even though there were few fellow hominids. *Evolution of man* timelines were pretty empty in those days. There just weren't many hominids beyond Piltdown Prank-man and Nebraska Tooth-man. Neanderthal man and Cro-Magnon had to be added to the mix to make it look like there was lineage. But both were humans, not human precursors.

(10) **Write Papers and Articles:** Of course lots were written. Science journals such as *Nature*, and *Scientific American*, picked up the story. Articles were printed in major newspapers all over the world.

So you see, *The Ten Steps to Hominid-ism* were already nearly fully evolved. The technique for forming hominids in support of the evo-illusion of man was already well developed in the early twentieth century.

Evo-illusionists circle the wagons to defend these uber-embarrassing evo-illusion moments. They make excuses for why paleoanthropologists were so fooled by Nebraska Tooth-man. There is no excuse. That anyone with a slight bit of intelligence could look at a tooth of any kind, and morph that into a North American ape-man is comical. There are no North American apes, except in zoos. Did they actually think this North American ape-man vanished into thin air because it evolved into modern humans? And it did so thousands of miles away from African and European apes, which were randomly doing the exact same thing... evolving into humans? Is that plausible? This wasn't a "scientific error". This was a perfect example of modern-day evo-illusionists, post

Aesop's fables, conjuring incredible scientific illusions. If the evidence for human evolution were plentiful, as it should be, there would be no need for illusions. Bones of pre-humans should be as common as fossil fish that are hundreds of millions of years older than humans. Fossils of pre-humans should be so common, one should be able to go into a fossil shop, and see dozens for sale. They should be routine finds. Massive illusions like Piltdown Prank-man, and Nebraska Tooth-man exist because there just wasn't any evidence for the evolution of man from an early ape ancestor. If evolution were good science, illusions wouldn't be necessary. Evolution would die without its illusions. Human fossils aren't the only illusions used by evolution scientists. In Chapter 13 of my book *Evo-illusion* I describe the illusions used to fool audiences into believing birds such as hummingbirds and woodpeckers evolved from a dinosaur that looked like a small T. rex. In chapter 14 I describe the illusions used to fool audiences into believing 400,000 lb. 100 foot long blue whales evolved from little 20 lb. fox-like animals. Evolution doesn't have what it needs to prove itself as a science, so the illusions will continue. Most will succeed, because few people will question or show any skepticism in this venue.

John Wolf and James S. Mellett wrote an article discussing and defending the paleoanthropologists who were part of the embarrassing Ape *Man of the Western World* illusion. Their final summation:

But what creationists ridicule as guesswork, and trial and error, and flip-flopping from theory to theory are the very essence of science, the stuff of science. Error correction is part of the creative element in the advance of science, and when disagreement occurs, it means not that science is in trouble but that errors are being corrected and scientific advances being made... We cannot conceive of two more diametrically opposed methods of explaining the world around us. One uses the correction of error as an inherent part of the process of searching for the truth, or ultimate reality in nature; the other rejects error or cannot admit its existence. Although it may be human to make mistakes, it is scientific to correct them. That is the nub of the issue between creationism and science.[24]

They couldn't be more wrong. The Piltdown Prank-man and Nebraska Tooth-man illusions aren't examples of trial and error in science. They're examples of the massive frauds direly needed by evo-illusionists to fool audiences into believing evolution is valid science. It's an example of how audiences are so gullible and desirous of falling for the illusions, they will believe anything presented by evo-illusionists. Evolution expert Heather Scoville said,

Maybe the word "hoax" is a bit harsh to use in the case of Nebraska Man, because it was more of a case of mistaken identity than an all out fraud like the Piltdown Man turned out to be.[26]

Making a single and obvious pig's tooth found in the dirt in Nebraska appear to be an entire ape-person is "mistaken identity"? Nebraska Tooth-man is an example of the idiotic lengths evo-illusionists will go to prove their illusions. See how, when an original illusion is uncovered, illusions are conjured to cover for the failure of the first illusion? I can't even imagine what would get into a person's mind that allowed him to think he could fool so many people with just a single pig's tooth. But it did get in Osborn's mind, and he and his co-illusionists did fool just about everyone in their audiences. Heather Scoville herself was either severely fooled by the illusion, or she made her own sub-illusion to cover for the immense failure of the Nebraska Tooth-man illusion. When the Piltdown Prank-man and Nebraska Tooth-man illusions collapsed, the excuses poured out. The excuses themselves are evo-illusions, and they become part of the major illusion of human evolution. The fact that these illusionists could fool so many people and scientists with such miniscule props is astounding. It proves audiences will believe anything, as long as it fits their overall belief system. We are no different than the ancients. We still do have a foot in the Dark Ages. I'm certain that most people who read the headlines in the New York Times about an entire ape-man being made to appear from just a tooth found in the Nebraska dirt accepted the news as fact. Most audiences think modern scientists are completely infallible. They think those ancient days when people were superstitious are laughable. Today we do *real* science. But do we? Or do we make up fables just like the ancients?

As of 1920, a half of a century after Darwin published his book *The Descent of Man*, that described the illusion that man evolved from earlier simpler forms, there were six main props as verification. Two, Cro-Magnon, and Neanderthal, were both already humans, and represented no evolution. They were used as evidence anyway. Remember, beggars cannot be choosers; and evo-illusionists were in the position of being beggars. Plus, they had one gross put-on, Piltdown Prank-man that had not yet been uncovered as a prank; and an immense mistake, a pig's tooth found in Nebraska and declared to be Nebraska man. Rounding out the evidence were Java man, a tooth and skullcap, and a human jaw found in Germany. I will discuss these two in the next chapter. Evolution was in an incredible state of disarray and panic. If *anyone* could dig up *anything*, the evo-illusion community, without the slightest bit of challenge, would jump, and accept it as evidence.

The next few chapters will describe the evo-illusions that followed up the Piltdown and Nebraska fiascoes. Maybe just a pig's tooth isn't quite enough to fool everyone. But if someone could come up with just a teensy bit more stuff, would that then be enough? You will soon to find out.

Chapter 3

China, Germany, and Indonesia Get Into the Evo-Illusion Business

It's far easier to fool someone than it is to convince them they have been fooled. -Mark Twain

Piltdown Prank-man, and Nebraska Tooth-man should have embarrassed the heck out of the world's evo-illusionists. What could be more humiliating than being caught supporting not one, but two immense deceptions? So were Nebraska Tooth and Piltdown Prank the only hoaxes of their day? Of course not. There are newer evo-illusions that managed to escape careful scrutiny. They live on as the modern version of human evolution. With so many scientists willing to put their reputations on the line, and to continue supporting the illusion of human evolution, these illusions have so far been able to escape the humiliation of being exposed. In fact, original miniscule and completely illogical finds declared to be human precursors were added to and built on by other finds in other locations. If you can conjure a hominid illusion, and get thousands of other evo-illusionists to treat it as valid science, and even to add their dug-up bones to the evo-illusion, it becomes valid science; at least in the minds of evo-illusion's supporters and believers.

In this chapter I will discuss three illusions of human evolution that escaped the scrutiny that Nebraska Tooth-man and Piltdown Prank-man suffered. They survive today in documentaries, textbooks, and museums throughout the world as examples of accepted human ancestors. Are they any different than Piltdown Prank-man and Nebraska Tooth-man? I will let you decide.

The *Java Man* Illusion: In 1891 paleontologist Eugene Dubois (Figure 3-1) unearthed a broken cranium of an animal on the Indonesian island of Java. Dubois had enlisted as a doctor with the Dutch medical corps in the Dutch East Indies. For some strange reason known only

Fig. 3-1

to him, he thought there might be missing link fossils there. He planned to dig in his spare time. Convict laborers and two army sergeants assisted him with his dig. Since even Darwin thought humans evolved from a lineage of apes in Africa, still considered to be valid theory, why on Earth did Dubois think it might be a good idea to search for human precursors in the millions of acres of dirt on an island that early hominids could not possibly have migrated to? Dubois picked the strangest and most unlikely location to dig. Try to put your mind in the mindset of the time. The fact was that virtually no hominid fossil bones were found anywhere on the entire planet Earth, even though there were numerous frantic and detailed searches. The only evidence Dubois had to go by was writings such as Darwin's book, *The Descent of Man,* which said humans came from Africa. So what did Dubois know that made him want to go to Java to dig? Why the island of Java, when there were many other enormous landmasses nearby that had an immensely greater chance of success? Why didn't he pick Africa for his digs, the supposed birthplace of humanity? Evolution is filled with absurdities that have miraculous outcomes, such as this one.

Dubois first digs were along a riverbed. His completely illogical notion to dig in Java was quickly rewarded. The team soon excavated a molar and a skullcap. What an unbelievable stroke of luck. What a very strange combination: a tooth and a skullcap, and no more? Did the rest of the teeth and skeleton vanish without leaving a trace? The tooth and skullcap became known as *Java man.* Apparently a group of pre-humans did migrate thousands of miles, all that way from Africa to Java, over 700,000 years ago. One individual out of the entire group of Java men left a tooth and skullcap for Dubois to find 700,000 years later. This amazing story would be more believable if fossil bones that match those found by Dubois were numerous and located on a path from Africa, through Asia, *and* on Java. How could only one partial skull and a tooth be all that was left from a population of Java people? What happened to all the other remnant bones?

G.H.R. von Koenigswald helped Dubois a tiny bit, 45 years later. In 1937 he unearthed a very similar and slightly more complete braincase at Sangiran, also in Java. The volume of Dubois skullcap was 940 cc. G.H.R. von Koenigswald's was even smaller, at only 815 cc. (Humans have a brain volume of about 1400 cc.) Besides this second find, no additional hominid skeletal parts were dug up anywhere else on Java either accidentally, or by diggers looking for human fossils. A group of German paleoanthropologists went to Java in 1907 in hopes of finding more hominid fossils. They hired 75 diggers.

They dug on the same sedimentary level as was the one that turned up Java man's skullcap. All they found were remnants of modern animals. There were no other 700,000-year-old hominid parts, or parts from hominids of any age or type. This makes Dubois' bull's-eye/hole-in-one find exponentially more improbable. It seems there is a much larger question here than whether Java man is ape or human. It involves the validity of the finds. Were the Java man bones planted fakes concocted by Dubois and von Koenigswald or their diggers? The odds of finding two of the exact same skeletal parts, partial craniums, from two different pre-humans, *and nothing more*, for almost one hundred years is, well, incalculable. Also, the fact that in 700,000 years, no other hominid remnants were left is beyond astounding; a real deal breaker.

A year after his skullcap find, Dubois uncovered a human thighbone 40 feet away from his original find. Was it related to the skullcap? Is there any way of knowing for certain or proving that it was? Dubois and von Koenigswald could argue that only skullcaps survived because they're composed of such thick bone. That argument is invalid because the thighbone Dubois claimed was from his fossil find survived just fine. Additionally, the skullcap was fully intact. It didn't show signs of being eroded or dissolved. It was a full-thickness skullcap. So where are Java man's other skeletal parts; his long bones, his ribs, and pelvis, and the rest of his skull? How could 100 percent of the remainder of the skeleton vanish? Shouldn't there be many remnants, or at least some eroded and thinned out bone chips nearby? Why would only one long bone, one tooth, and one partial skullcap survive intact, while not even a tiny part of any other bones did? Where are the long bones from von Koenigswald's find? How is it possible that no other skeletal part from Java man survived? There is only one plausible explanation. The skullcaps were planted at the dig sites from some other location. Ignoring Dubois' tooth and femur, this is the only possible explanation for the finding of one single skullcap in each of two locations, and nothing else in either location. The probability math is basic. It's a possibility that is never explored by any evo-illusionist. Almost every find of any bone is taken to be a serious hominid find. Other possibilities are rarely considered.[1-3] One exception was Dr. Rudolph Virchow, Director of the Berlin Society for Anthropology and founder of the science of pathology. He examined Dubois' fossils and wrote:

The skull has a deep suture between the low vault and the upper edge of the orbits. Such a suture is found only in apes, not in man. Thus the skull must belong to an ape. In my opinion this creature was an animal, a giant gibbon in fact. The thighbone has not the slightest connection with the skull.

The sutures were ape-like? Sutures are seams in bone where two parts meet during embryonic development. So, you see, there are other characteristics that can be utilized to delineate an ape skull from a human skull than the ones I cited in Chapter 1. These characteristics say Dubois' find was ape. [4,5]

Two views of each of the three fossils are pictured in Figure 3-2. The skullcap had a large boney browridge, a small flat cranium, which terminates inside the brow ridges (arrows), and no vertical forehead, which of course makes it an ape. There were arguments other than Virchow's about whether the thighbone, which is modern human-like, belonged to the skull. Of course there is no way of proving it did. It could have come from other sedimentary layers, and simply landed near the

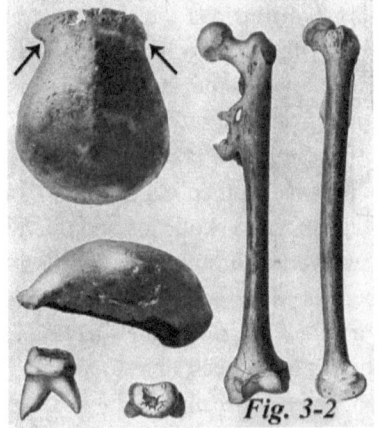

Fig. 3-2

skullcap. Bodies of humans are frequently found floating down rivers in third world countries such as Indonesia and India. Or it could have been purposefully placed, like Piltdown Prank-man's bones were. Despite Dr. Virchow's analysis, despite all logic, and the fact that the thighbone was found a year later and forty feet away, Dubois credited the thighbone as being from the same skeleton as the skullcap. This would support the illusion that Java man was an upright walker. Again, I find it rather amazing that zero of 625 modern primate species make tools, and zero walk upright. This is just too perfect; way too perfect to support human evolution. Even so, the Java man illusion was hatched. Dubois' single, unattached thighbone, found forty feet away and a year later, made Java man an upright walker; no question about it.

Fig.3-3

The skullcap, the dark part of the Java man skull in Figure 3-3, was bestowed the complex name *Pithecanthropus erectus*. The skullcap was used to concoct the entire skull and full model shown in the photo. It's interesting that, if an untrained person looked at the skull, they would think the find was the white portion of the skull. Or, they might think the entire skull was the actual find. When I was in school, I was taught the model of Pithecanthropus erectus was a human precursor. I had no idea that it was made from just a skullcap. I believed what I was told and was shown. I look back now and am amazed at how the evo-illusions that I was taught in school had such an effect on me. I believed without question or skepticism. Why wouldn't I? I was amazed at how far science had come; how genius model makers were able to construct entire hominids from just a few boney parts; or just a few teeth. It never entered my mind that they were fakes.

Here is another massive part of the illusion: the notion that the mere existence of these skullcaps and the few other bones means for certain that Java Man evolved into humans. What is it that makes this illusion valid, when it's nothing but pure fantasy? Even if these bone fragments have any similarity to humans does not mean they have anything to do with modern humans. Evo-illusionists have been able to fool people into thinking the mere existence of any part of a skull makes the source pre-human. Five years before his death, Dubois admitted his skullcap was from a large gibbon.[6-9]

Here is what Dubois said, in a paper published in 1935:

Pithecanthropus [Java man] **was not a man, but a gigantic genus allied to the gibbons**, *however superior to the gibbons on account of its exceedingly large brain volume and distinguished at the same time by its faculty of assuming an erect attitude and gait. It had the double cephalization* [ratio of brain size to body size] *of the anthropoid apes in general and half that of man.*[10]

How on Earth would Dubois know the ratio of brain size to body size, when all he had was a skullcap of some kind of ape? Even if his dubious thighbone did somehow match the skullcap, there was no real way to determine body weight and volume. In a paper he wrote only two years later, in 1937, Dubois said;

It was the surprising volume of the brain - which is very much too large for an anthropoid ape, and which is small compared with the average, though not smaller than the smallest human brain - that led to the now almost general view that the "Ape Man" of Trinil, Java was really a primitive Man. Morphologically, however, the calvaria [skullcap] *closely resembles that of anthropoid apes, especially the gibbon...* ***I still believe, now more firmly than ever, that the Pithecanthropus of Trinil is the real 'missing link'.***[10]

The big question is, what was Dubois smokin'? What a complete turnaround. He believes "now more firmly than ever"? Did one of the other evo-illusionists get hold of Dubois and scold the heck out of him for not promoting his illusion? If Java man were just a gibbon, it would be worthless. If it was a human precursor, the missing link, its value would rise thousands of times, and so would Dubois' value in the world of evo-illusion. What on Earth gives the person that dug up Java man the power to determine if it's an ape or a human or the missing link in the first place? He should be the very *last* person who would have a say, as his bias would obviously overpower his logic. But Dubois was an evo-illusionist. As such, he can say anything to his audience, who wants to be fooled. The motivation of any person in Dubois situation would be fame and money, of course. Several articles argued whether or not Dubois thought his find was a hominid. It seems strange that other evo-illusionists tout that Dubois's opinion had such great importance. But, as the finder evo-illusionist, he can say anything he wants. Dubois obviously had great incentive and motivation to give his skullcap the title of "missing link". And evo-illusionists had great motivation to have Dubois declare Java man the missing link so they could continue with their own illusions. Dubois' turnaround was a win-win for him and all scientific illusionists who promoted the illusion of human evolution.

The only true verifiable evolution known to mankind is performed by evo-artists and evo-sculptors, who are key to making this whole illusion so believable. Evo-artists and evo-sculptors began doing their magic on Java man practically as soon as the skullcaps were found. They morphed a couple of skullcaps into full early human beings.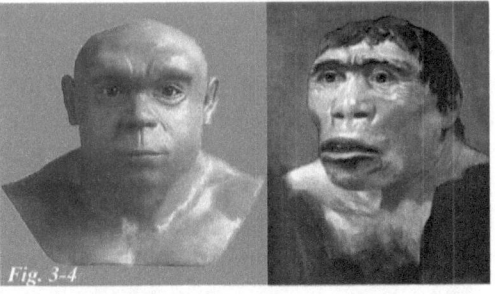

Fig. 3-4

Figure 3-4 shows two examples of what evo-sculptors and evo-artists can do with just a tooth and a broken piece of skull. They even made human skin and expression appear out of nowhere. Why didn't they make Java man's imaginary head look like that of a gibbon, complete with the fur primates have that covers their heads? Well, that would kill the ape-evolving-into-human illusion, of course. I wonder where the guy in the painting on the right went for his haircuts. They must have had barbers with rock scissors then? When you look at these pictures, realize you are looking at pure illusions.

Like the discoverer of Piltdown man, Dubois was grandly rewarded. In 1897, the University of Amsterdam conferred Dubois an honorary doctorate in botany and zoology. He was given a full professorship in 1899. For thirty years after his find he was *the keeper of paleontology, geology and mineralogy* at Teylers Museum, where he also kept the H. erectus remains. See what happens when you find a fake fossil? The pats on the back will be enormous.

Germany takes its turn: A major find and a new illusion was conjured when, in October of 1907, Daniel Hartmann, a gravel pit worker, walked into a pub in a village near Heidelberg, Germany, and announced that during his digging that day he had found *Adam*. What he had found was an almost complete, fairly robust human jaw that appeared to be very old. (Figure 3-5) Paleoanthropologist Otto Schoetensack from the University of Heidelberg was contacted the following day. Schoetensack was of course very excited because this jaw *must* represent a species of ancient pre-human. I

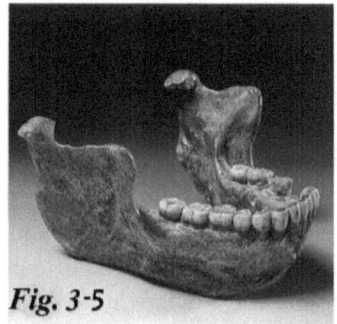

Fig. 3-5

mean, what else could it possibly be? How could he tell so quickly, from just a jawbone? Well, he did. He even gave the hominid owner of the jaw its complex name, as all good evo-illusionists do. He called the jaw *Homo heidelbergensis*. As I stated earlier, evo-illusions work well on audiences that are shown skullcaps and jaws that have been bestowed with technical hominid names. Technical names are a key step in the birthing of the illusion that a bone fragment came from a full human precursor. Yes, in only a few short days a jaw found in a gravel pit turned into a fully formed pre-human. Now that's rapid evolution. If a single pig's tooth can be turned into a human precursor, as was the case with Nebraska Tooth, turning a jaw into a human ancestor should be a cinch. My question is, what on Earth would make an uneducated gravel pit worker think he had found Adam, or a human precursor? Daniel must have been smart beyond imagination. Most people would have taken it home and put it in their sock drawer; or at best, called the police, wondering if it might be evidence in a missing person investigation. But no. This gravel pit worker thought it would be important to take the jawbone to a paleoanthropologist. Which he did, and the rest is history. Another ape-man illusion is born.

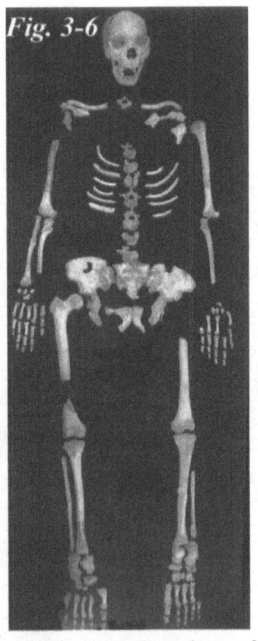

Fig. 3-6

Heidelberg man became one of the most important finds in the history of evo-illusion. That single jaw turned into the entire skeleton in Figure 3-6. How did this happen? No further remains of Homo heidelbergensis were discovered at the Heidelberg site despite extensive digs and searches. How strange that only a jawbone was found. What could have happened to the rest of the poor bloke that owned that jaw? That must be a whole story unto itself. This is the *Heidelberg 99 percent missing skeleton remnants mystery,* which is just like the incredible *Java man 99 percent missing skeleton remnants mystery.*

So how did a single jaw morph itself into the full skeleton in the photograph? Astounding, as it may seem, bones were taken from all over the world, and pieced together to make this skeleton. Long bones dug out in one location of the world were added to ribs found in another. And, like a puzzle, but with pieces that don't fit, it was gradually assembled. Notice that none of the ball and socket joints articulate. Heidelberg man fooled audiences, one piece at a time. This is how really good evo-illusions are formed. Most people looking at a picture of the skeleton, if they don't read the fine print, if the fine print were even there to read, would think Homo heidelbergensis was dug up as a complete skeleton, in one location. Most people who view the Heidelberg skeleton, if they were aware that evo-illusionists put it together from bones found at distant multiple sites, would accept it as valid, without a bit of skepticism. After all, *real scientists* did the assembling. Anything real scientists do is real science. Right? The illusion that the source of the Heidelberg skeleton was from one single hominid goes unchallenged. Few people who look at Mr. Heidelberg would have any idea they're looking at an illusion. [11,14]

Here is a partial list of the bones used to create the illusion that Heidelberg man is a full skeleton:

(1) The Mauer mandible, found in 1907, near Heidelberg, Germany. Age, 700,000 to 400,00 years old. Found by gravel pit workers.

(2) Petralona 1: facial bones with a partial braincase. Found in 1969 in Petralona, Greece. Its age is about 500,000 to 250,000 years old.

(3) Tautavel Man: fairly complete facial bones, five molars, part of the braincase. Found in 1971 in Arago, France. Its age is about 400,000 years.

(4) Kabwe man, a nearly complete cranium, jaw fragment, sacrum, portions of a pelvis. Found in 1921 in Kabwe, Zambia. Its age is about 200,000 to 125,000 years old.

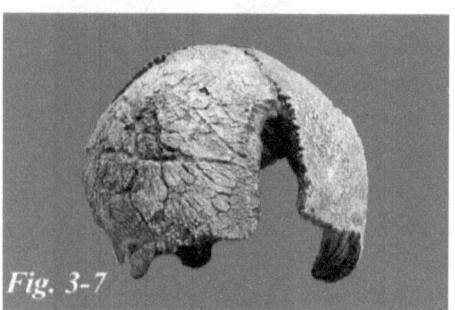

Fig. 3-7

(5) Tibia (shin bone) found in Boxgrove, West Sussex, UK

(6) Three more broken sections of a skull are from Swanscombe in Kent, England, found in 1935, 1936 and 1955. Figure 3-7 is a photo of the 1935 discovery.

These skeletal parts are from Greece, England, Africa, and Germany; locations thousands of miles apart, and hundreds of thousands of years apart in time. There is no possible way that hominids from Africa populated all of these distant locations. Notice the youngest bones are from Africa, the exact opposite of what evo-illusionists say happened with human migration. The oldest should be African, not the youngest.

The rest of the skeleton pictured is composed of bones found in other locations throughout the world. Heidelberg man was truly an international playboy; a Renaissance man. [11-14]

Guess what this guy in Figure 3-8 is doing. He's looking for bones from Heidelberg man in a tiny cave in France. Yes, France. Actually this picture was placed in a textbook on evolution called *Evolution: The Human Story* to fool readers into thinking there are people

Fig. 3-8

like him all over the world who are constantly digging for bones of hominids. This picture is actually just a sub-illusion, another part of the illusion of human evolution. It creates the illusion that makes the audience think new finds will constantly be made. They think that as a result of dedicated diggers like this guy, they will get more and more information about the evolution of mankind as the years go by. Information that the audience gets from finds will allow them to understand how humans evolved from an early ape ancestor. Real science will march on. If you really look at the massive areas where humans supposedly evolved, and the miniscule number of bones that have been found, digging for human precursor bones is a colossal waste of time and money. Who pays the diggers to dig? Who pays the paleoanthropologists to oversee the diggers? Looking at the results of the digs, I doubt there are many grants that would pay people to go to Africa, or Asia, or Europe and search around and dig when there is an infinitesimally small chance of anything showing up. The guy in this picture probably spent years in school getting his PhD in paleoanthropology... and here he is, doing what he always dreamed of. All I can say is it sure is a cute picture. I wonder why he isn't looking at the camera. Don't most people look at the camera and smile when they're getting their picture taken? I also wonder if he found any bones. I sincerely do hope he did. I also hope that nice set of well-fitting gloves prevented him from getting blisters, and dirt under his fingernails. I wonder who pays him. Hey, maybe the Smithsonian National Museum of Natural History, that says:

We don't know everything about early humans—but we keep learning more! Paleoanthropologists are constantly in the field, excavating new areas with groundbreaking technology, and continually filling in some of the gaps about our understanding of human evolution.[15]

You see, along with pictures like this, and the notion that diggers are constantly in the field, digging up new hominid finds, the illusion of human evolution is enhanced. Just think, the revered Smithsonian is part of the illusion. If diggers were constantly digging, there would constantly be new finds. New and very questionable hominid finds show up about every ten to twenty years or so. Which means diggers aren't in the field, constantly digging, or they sure aren't finding much for all of their efforts. Can you imagine digging in dirt in Africa for five or ten years and getting basically... nothing? How about spending ten years in caves like the guy above, and getting... not much. Who would devote their lives to that kind of frustration and uber-boredom? What is funny to me now is that, when I was an evolution believer, I always thought it would be great fun to dig for fossils in Africa. I have since given that notion a huge mind change.

From the single Heidelberg jaw, it was determined that Heidelberg man migrated out of Africa, and lived in Europe from about 700,000 to 200,000 years ago. Interesting, because man didn't migrate out of Africa until 60,000 years ago. Isn't modern science wonderful? Just think, they came to this astounding conclusion... from just a jaw.

Here is where it gets even more amazing and difficult for evo-illusionists. They need to describe how random mutations that would be exponentially different in each geographic area that Heidelberg man and Java man migrated to all resulted in identical outcomes; they all evolved into Homo sapiens. All of the groups of descendants of Heidelberg man who lived in vastly different locations had to evolve large flat vertical foreheads, large ovoid craniums, flattened bony brows, and receded flattened jaws. Along with all of these, they had to evolve human-type skin, and their brains had to evolve large frontal lobes and other brain parts that would result in human intelligence and consciousness. Even stranger, all groups of hominids evolved the ability to procreate with every other group. They all evolved into the same species. Each geographically isolated group had to do this completely independent of each other geographically isolated group.

From that one single jaw found in Heidelberg, evo-illusionists gleaned an enormous amount of information. They like to use the terms *perhaps, probably, maybe, might have, could have, and may have* to concoct their illusions. The following is a conglomeration of only some of the amazing statements that I found in evo-illusionist papers about Homo heidelbergensis that were gleaned from just a jawbone:

Heidelberg man was *probably* the first early human species to live in colder climates.

Perhaps Heidelberg man built controlled fires to keep warm.

Perhaps this early human also broke new ground; maybe it was the first species to build shelters.

Perhaps he created simple dwellings out of wood and rock.

Animal hide clothing *may have* been worn by Heidelberg men and women; especially by populations living in the cooler European areas.

Their short, wide bodies were *perhaps* an adaptation to conserving heat.

But their short wide bodies were complete illusions, since all that was found was a jaw. So the "colder climates adaptation" notion is an illusion as well. I guess maybe their lower jaw didn't conserve heat; it was so big. But the rest of their non-existent bodies did. One wonders why they left Africa, with its ideal climate, to venture to freezing cold Heidelberg. I bet they left Africa in the summer, and ventured to Heidelberg, not realizing how cold the winters would be. When winter hit, boy their leader must have been in a world of trouble. Can you imagine how pissed, and freezing cold the Heidelberg people must have been?

What is really astounding is that eight wooden spears, called *Schöningen Spears*, (Figure 3-9) were found near the site of the Heidelberg jaw find. Even the spears were blessed with a complex name, which adds to the illusion. *Perhaps* these spears were made by Heidelberg men? Well, according to the best evo-illusionists: yes they were. These spears are over 400,000 years old, and are considered the oldest completely preserved hunting weapons in the world. I guess we could say they truly are antiques. They're regarded as *perhaps* the first evidence of hunting by Heidelberg man. There is just one tiny problem here. Does wood last 400,000 years anywhere in the world? I just think of my wood patio cover, and how long it lasted. After 30 years, moisture, bugs, fungus, and bacteria destroyed it. I don't think the remnants of this spear would have lasted 100 years, much less 400,000 years. All of Mr. Heidelberg's bones vanished or dissolved or something, except for his jaw. But his wooden spears lasted 400,000 years, fully intact? Right. This illusion is getting out of hand. Can it be proved that Heidelberg man made those spears, and not some group much later in the 400,000-year time span? Like maybe fifty years ago; max? Anyway, all of this information was gleaned from just one jaw, and lots of *perhaps's*. [16,17]

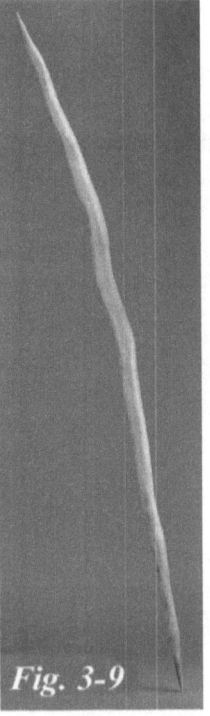
Fig. 3-9

According to evo-illusionists, the average size of Heidelberg man was about 5' 9" and 136 lbs. The ladies averaged 5" 2" and about 112 lbs. How did these statistics come to be, when all that was found was one jaw at Heidelberg? What numbers did evo-illusionists use to get their average? I'd love to see their calculations. Using a single jaw from a single person to make an average height and weight of males and females of a population would get an "F" grade in any statistics class at any university in the world. This is what would be the epitome of *anecdotal evidence*. In case you're not familiar, anecdotal evidence is the use of one singled out example to determine trends and characteristics of entire populations. Evolution's illusions of human evolution are made up completely of anecdotal evidence, as you have seen, and will see.

Evo-illusionists that tout Heidelberg man as our ancestor are a bit insecure in their theory. So they're diligently and patiently searching for further fragments at known discovery sites in the assumption that these sites hold more fascinating fossils. *The Isotope Geochemistry Research Group* led by Dr. Bernd Kober from the Institute of Earth Sciences at the University of Heidelberg is using a state-of-the-art *thermion mass spectrometer* to analyze the Heidelberg sand pit where Daniel Hartmann discovered his Adam in 1907. Just think if he came up with another bone, say, a skullcap. That would really mean a lot. Wouldn't it? I for one am really excited about other possible finds. Who knows how many more illusions might be conjured. [18-21]

Fig. 3-10

Well, I can't leave Heidelberg man without showing the only true and proven evolution in existence. Of course I'm talking about the illusion of human evolution fabricated by evo-artists. The rogue's gallery (Figure 3-10) is a collection of five different versions of Heidelberg men. All five evo-artists used only the Heidelberg jaw as a model to construct these hominids; or maybe they used the conglomeration of skull parts from all over the world? I wonder if they even looked at any bones at all. Heck, they didn't need them since they model and paint not for accuracy, but to make the anthropologists and museums that pay them happy, just like any paid employees. I guess you could say these evo-artists are inventive geniuses. They gave the Heidelberg men human expressions, human skin, and human hair, human eyes, human beards, human foreheads... Again, I wonder how Heidelberg men cut their hair. They must have made scissors and razors out of stone. Evo-artists make action paintings of these guys. They're shown hunting with spears and stone weapons, and sitting around a campfire, and using their stone cutting tools, while the gals cook and nurse their babies. Actually, it really doesn't matter since they're all imaginary, and modeled after a human jawbone. The jaw is definitely not protrusive. The chin is shaped like a human jaw, so it's original owner and his friends may simply have been humans. Which means they certainly aren't 700,000 years old, since evo-illusionists say true humans first appeared 200,000 years ago. Whoever set Heidelberg man's age at 700,000 years might want to go back and recalculate. Isn't it strange that Eve was made from Adam's rib, and Heidelberg man was made from a jawbone? How coincidental.

A man from Peking named Sum Ting Wong: A third fossil that was inexorably tied to the Java man and Heidelberg man illusions was found thousands of miles away from each in China. You see, an evo-illusionist may perform an illusion that doesn't quite fool an audience, which may have been the case with the Java man illusion. Some people might have been skeptical. If the other evo-illusionists can then successfully perform a similar illusion a second time, and then a third, the whole audience will be dumbfounded. Even the strongest skeptics will be fooled as I was. The new find was given the name *Peking man.* It's an even more astonishing illusion than Java man and Heidelberg man. Workers digging in limestone mines at Zhoukoudiann near Peking in 1917 uncovered sedimentary strata. Some found what they called "dragon bones", which were actually fossils of ancient animal species. The animal fossils attracted a number of fossil hunters. For some strange reason, they all were hoping beyond hope that they would be the first to dig up a true "missing link". Again, one might wonder why so many would be attracted to any area of China, since it was already known that modern humans first originated in Africa, and migrated from there to populate the world. What evolution had, up until 1917, for a missing link was incredibly piecemeal, dubious, and sketchy. Remember in the thirty-five years since Darwin's death, all that was discovered to support his theory of human evolution was the Cro-Magnon and Neanderthal humans, Piltdown Prank-man, the Heidelberg human jaw, and the very questionable Java man. Darwin's ideas about the evolution of man were really without evidence. The person who came up with the first true missing link, a real skeleton with corroborating evidence, would be declared an evo-hero, and be grandly rewarded and recognized by evo-illusionists all over the world.

Johann Gunnar Andersson, a well respected Swedish geologist and archaeologist, who, serving as an adviser on mineral affairs in the Ministry of Agriculture and Commerce of the Chinese Government, had been digging in the sediment at Zhoukoudiann for four years in hopes of finding the missing link. For four years he had no interesting results. Can you imagine digging at a site for four years, with no results? Why would anyone do that? Particularly in China where there wasn't supposed to be any hominids.

Figure 3-11 is a photo of the site where all of the digging took place. Pretty alluring, eh? Morbid boredom would have set in if I had to go there and dig on a regular basis. How on Earth did they get

Fig. 3-11

anyone to sign up for digging in that mess? In 1921, after all that time and extensive digging, Andersson and Otto Zdansky, an Austrian paleontologist who teamed up with Andersson, hit the jackpot. They discovered two human-like teeth. One was an adult upper molar the other was an unerupted lower child's tooth. Can you imagine the celebrations that must have ensued? I bet they were frolicking in the bars in Peking that very night. I bet each guy kept one tooth in his pocket. I mean, how could they trust each other with such valuable finds. Just think, those two teeth held clues to the origin of the entire human species. What a responsibility.

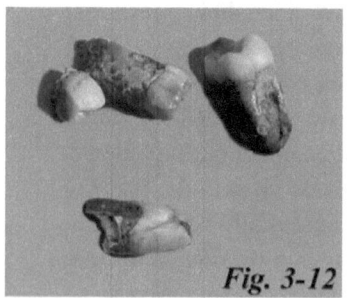

Fig. 3-12

The next year they found two more teeth, for a total of four teeth. All four teeth are pictured in Figure 3-12. That's an average of one tooth per year. I'm certain the additional teeth brought more celebrations. Wouldn't you be happy if you dug for four or five years, and finally found a couple of teeth? One article said the news of the finds "*astonished* the scientific world since at that time there had not been any discovery of any such ancient human fossils in China or any other country in Asia." I say, these four teeth are human fossils? Well, of course, with all of this excitement, other anthropologists joined the team. I mean what could attract a legion of fossil hunters any more than four teeth found in Chinese dirt? Dr. Davidson Black, a Canadian anthropologist and Dean of the Anatomy Department of Peking Union Medical College, "thoroughly studied the upper molar" found in 1921. Being a dentist myself, I love the lingo, "thoroughly studied". It's made to sound like the "thorough study" was a major undertaking. How long does it take to "thoroughly study" a tooth? It would take me only a few moments to tell if it was of human origin. The only molar in all photos I found was a *lower* molar. (Upper right in Figure 3-12) Dr. Black should have taken a few more classes on dental anatomy. Well, now that four teeth were found, the thorough studier of the teeth did the obvious thing. He determined the teeth were *Hominidae*, a new genus, and an astounding determination. Just think, a whole new genus magically appeared from only four teeth. The new hominid composed of four teeth was bestowed the complex name, *Sinanthropus pekinensis,* which automatically gave it credibility and made it a valid human precursor. See how scientific illusion works? Four teeth evolved into a full early human being, with a complex name, and a new genus, in months, not millions of years. So the first three steps in forming the illusion of human evolution has been accomplished. The four-tooth find now has its complex name, *Sinanthropus pekinensis*; and its cute name: *Peking man.*

Once the determination was made that the four Peking teeth were from a human ancestor because of "thorough study", the funds poured in. Evo-artists and evo-sculptors took over and concocted models and paintings of what Peking man "really" looked like. (Figure 3-13) The Rockefeller Foundation then enthusiastically funded further research on Peking man's teeth and Peking man himself, who of course was invisible. The illusion even fooled the members of the foundation. Can you imagine the discussion by the board of directors? It probably

Fig. 3-13

went something like this:

Gentlemen, there has been some astounding discoveries that prove that man came from ape! Yes, four teeth! In China! This is the most exciting scientific news of the millennium! It's been determined that those four teeth are from the missing link! The link's name is Sinanthropus pekinensis or Peking man; he walked upright, made tools, and all kinds of other stuff. What do you say we send whoever found those teeth a few hundred thousand dollars and see if they can dig up any more stuff! All those in favor say Aye!   ...Aye! Aye! Aye!..

With all of the new funding, extensive systematic excavation of Zhoukoudian Site was undertaken. These were exciting times. What really astounds me is that scientists in the 1920's were able to simply look at a tooth and dub the owner of said tooth a pre-human ancestor. Did those teeth truly belong to an ancient owner? Or maybe there was a modern human owner, who was one of the diggers, with a bad tooth condition and drool problem. The

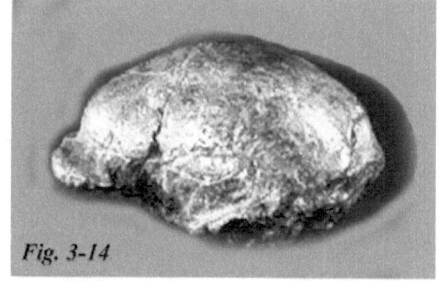

Fig. 3-14

diggers, and most people living in that area at that time, had horrible teeth. When teeth become as badly decayed as that cuspid, the bottom of the four teeth in the picture, and when teeth get severe bone disease called *periodontal disease*, they can simply fall out of their sockets, or be easily and painlessly lifted out with the fingers, with no help from a dentist whatsoever. It wasn't an uncommon occurrence at all. In the 1920's, nearly every adult human was edentulous (toothless) by the age of 50. There were endless supplies of extracted teeth in China in the 1920's. If I had to put odds that the teeth were either (a) purposefully or accidentally placed by a modern human, or that (b) they came from a hominid hundreds of thousands of years old who ventured to China from Africa, and who later evolved into modern humans. I would bet on (a) by billions to one. The child's tooth isn't an unerupted tooth at all. It's a normal primary tooth that had its root resorbed (dissolved) by an underlying erupting adult tooth. It simply fell out of a child's socket because its root was gone, normal fair with children's teeth. An *unerupted* tooth would have a full root. Apparently the finders didn't consult a dentist, and Davidson Black wasn't knowledgeable about teeth. A dentist would have killed the illusion. Maybe that's why. What really amazes me is that this illusion fooled anyone. Is it real science? Is this the twentieth century? Are we really out of the Dark Ages? With the Peking man find, the illusion of the evolution of man was really on its way. Darwin would have been so excited. Maybe.

In 1928, Dr. C. C. Young, a respected Chinese paleontologist, and Wenzhong Pei, a young Chinese geologist, joined the Peking man excavation and uncovered two lower jaws that were assigned to Peking man. One year later Pei found an almost complete skullcap. (Figure 3-14) So, for seven years, all there was to Peking man was four teeth. Magically turning four teeth into an entire pre-human being is a far greater illusion than I have ever seen on television or on any Las Vegas stage. The only greater illusion was Nebraska Tooth-man: a whole hominid from a single pig's tooth. It fooled people all over the world. If Peking man's four teeth found earlier were not enough to convince everyone that they belong to a human precursor, the skullcap found seven years later did the job. It was much more convincing for paleoanthropologists who were universally starved for evidence of human evolution. Just imagine. The diggers dug for four years and finally found four teeth. They were so excited about the puny find, they continued digging for *seven more years*. They then found a skullcap. I wonder what their wives thought, if they had wives. I really don't know if they dug constantly, or returned to the dig on a regular basis. Either way, the fact that the diggers went back to Zhoukoudian to dig when only four teeth were found in over ten years means the illusion really fooled the diggers far more than it did anyone. Just imagine someone dedicating a decade of their life to looking for human precursors by digging in dirt that turned up only four teeth. Astounding.

More minor bones were discovered between 1929 and 1966. By this time the Peking man illusion was well established as a true missing link. The new finds included partial lower jaws, skullcaps and braincases. The jaws came from an entirely different layer, which was 85 feet higher than the original skullcap. All skulls had ape characteristics. The large brow ridges alone are a dead giveaway. The skulls are ape. The illusion is that they're man's precursor.

The excitement and overwhelming hunger for missing links in the 1920's was astonishing. People would jump at the chance to believe any illusion that, in their mind, proved Darwin's theory. Newspapers would write stories based on their writers excitement, with no corroboration whatsoever. Such was the case with Peking man. The funds put up by the Rockefeller Foundation were quickly rewarded, even though the reward was fleeting. The New Your Times wrote, on December 16, 1929,

*The discovery in a cave near Peking of the fossilized bones of **ten men, who possibly lived 1,000,000 years ago**, as reported by scientists representing the Rockefeller Foundation and the Geological Survey of China, is held here to excel in interest all previous findings of this kind. Of paramount importance is the discovery of a perfect skull, now in the possession of Dr. Davidson Black, a Canadian paleontologist, which, it is asserted, bears characteristics showing that even at the beginning of the ice age there existed **men with the power of thinking** and who, unlike the "ape men," **walked erect**. From the fact that **the*** *ten skeletons lay huddled together in the cave, found in a field at Chou Outien [sic], thirty miles from Peking, the scientists hold that they led a community life.*

Nature, a prominent science journal, was also fooled by the illusion, and joined in by saying "the fossilized fragments of *ten more* examples of *Sinanthropus*", and "remains of *ten individuals*" were found.[22] News of this new illusion spread far and wide. The skullcap, some teeth, and few other parts that were found in the dig were magically turned into ten individuals who had the power of thought, walked erect, and had community life. The audience's excitement was palpable. Undoubtedly the board of directors of the Rockefeller Foundation hungered for information about early man according to Darwin, and news stories like this did a great job of feeding their hunger. They must have been ecstatic with this news after allocating hundreds of thousands of dollars to a dig that had turned up only four teeth. But the "ten individuals" story and the notion that they "walked erect, were able to think, and were organized into communities", turned out to be just another part of this incredible evo-illusion.

On December 28th 1929, the paleontologists from the dig held a news conference to clear the air. The illusion had gone too far, and it was too big to support. No skeletons from ten people were displayed. What was shown was just the single skullcap of Peking man. Incredibly, even this find was enough to make huge news around the world. Peking man became a celebrity, even with the unbelievable reversal that, *presto-change-o*, changed ten individuals who could think, walk erect, and have community life, back into just a skullcap and

a few teeth. It was kind of like when Cinderella's carriage and horses turned back into a pumpkin and a few mice. It doesn't take much to excite a hungry audience eager to be fooled. Over the years effort has been made to determine the source of the story about "the bones of ten individuals", but that mystery has never been solved, just like the mystery of who the Piltdown prankster was. Maybe the source *was* the Piltdown prankster!

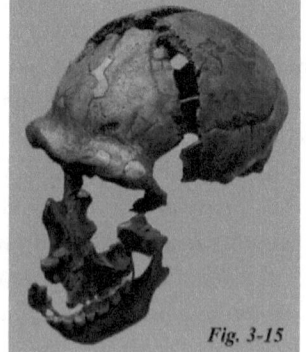
Fig. 3-15

Even with the shrinkage of the illusion from ten individuals back to not much more than a skullcap and four teeth, Peking man was now an established human precursor by agreement of numerous evo-illusionists. You see, illusionists themselves can be fooled by illusions as well.

Again, 99 percent of all species that have ever inhabited the Earth have gone extinct. Minimally, dozens of ape species must have also become extinct. Considering Peking man's flat skullcap, large brow ridges, small cranium, and prognathic jaws from another location, isn't it far more logical that he's nothing but an extinct ape, or maybe even a modern ape?

The "final" skullcap and jaws of Peking man (Figure 3-15) are actually a conglomeration of several finds from vastly different areas of the dig. Like Heidelberg man, Peking man is not the bones of just one "man". It's pure illusion that this picture is made up of bones from one single hominid head. Few people looking at this picture would have any idea that this Peking man skull was made up of bones from multiple individuals, from multiple locations, which greatly aids the illusion. Notice, as with Piltdown man, the condyles on the lower jaw have no fossa on the skull to fit into. What could possibly be the point of taking bones from different sites and different individuals, and putting them together to make them look like they're from one individual? This is a completely unscientific move purely designed to fool the audience. Most people looking at these fossils would be fooled, even if there were a sign below the bones in a museum "honestly" declaring that they are from multiple sources. Non-scientists would think that was OK because "real" scientists are doing the fooling. If scientists put these bones together, it must be scientific, so that's fine with them. The embarrassment of Piltdown Prank has not stopped evo-illusionists from putting bones together from vastly different sources, as it

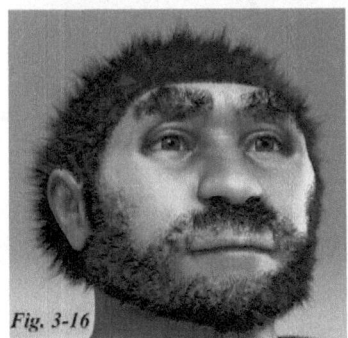

Fig. 3-16

should have. Every time bones are put together from different sources, the fossil evidence is poisoned. As long as the audience isn't aware of the poisoning, the illusion continues to fool.

From the few bones found at the Zhoukoudian Site, evo-artists conjured numerous complete Peking man hominids. The hominid pictured in Figure 3-16 was made to appear from the skull on the previous page, which was made from parts of four different skeletons found in different locations. Magically appearing on this hominid was human-type skin, a well coiffed beard, human eyes, human hair, and human expression. All of this out of a few teeth and bones. He looks like he's ready to go to make tools with his buddies. Or maybe shoot some pool?

The Peking man fossils had an interesting end. In the early stages of World War II, the Japanese were advancing on Peking. Many of the scientists at Peking Union Medical College, where the fossils were being stored, were seriously worried about the fate of their pathetic group of fossils. It was decided that they should be shipped to America for safekeeping. Shipping these fossils through a war zone on a ship that would take them thousands of miles to the United States seems like a pretty dubious plan, considering that none of the paleoanthropologists on this dig were American. One was Austrian, one was Swedish, and the rest were Chinese. So these guys were going to ship to the United States? Lucky for us, plaster casts were made of the fossils in case they were damaged or lost. Gosh, if they were lost in transit, how would we know where humans came from? The bones and teeth were packaged, and sent to the local shipping port at Chinwangtao. They were then packed in U.S. Marine Corps footlockers aboard the SS President Harrison. They were scheduled to leave China for the Philippines on December 8, 1941, the day after the United States declared war on Japan. But the ship ran aground at the mouth of the Yangtze River and was captured by the Japanese who used their newly claimed prize as a transport vessel. It was renamed Kachidoki Maru. The Kachidoki Maru was sunk in the China Sea by a U.S. submarine. The fossils were lost forever; well almost. [22-30]

There's an astounding twist to the mystery of the disappearing Peking man fossils. A unique discovery was recently made at the Museum of Evolution at Uppsala University in Sweden—a canine tooth from Peking man! (Figure 3-17) It was claimed that the tooth was "untouched since it was dug up in the 1920's in China". Per Ahlberg, professor of evolutionary developmental biology at Uppsala University, said of the find:

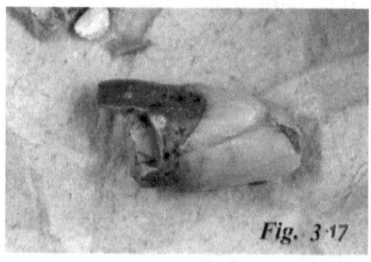

Fig. 3-17

This is an absolutely incredible find. Our Chinese colleagues and we are overwhelmed. With today's technology, a canine tooth that has not been handled can tell us so much more than in the past, such as what they ate.[32]

Such as "what they ate"? That's a "so much more" moment; from just checking out a canine tooth? Wow. Actually, since evo-illusionists can make an entire pre-human being materialize from just a few teeth, figuring out what they ate is chump change in comparison. I'm sure this tooth brought great celebration and festivities at the Museum of Evolution at Uppsala University. I bet the paleoanthropologists were painting Stockholm red that very night. They had just re-found the Peking man fossil! Just think, from this single tooth, evo-illusionists can support an entire human precursor, since all other bones and teeth are gone. They've done it before. Lots of faked excitement and articles about this find will be more than enough to keep the illusion of human evolution moving right along.

The big question is, is this find at Uppsala University *really* a tooth from Peking man? Or is it just another illusion? Why aren't the other teeth and bones with this one? Did they mysteriously travel to other obscure universities? How did they get separated? How would one prove that this definitely is the original find? Did Johann Gunnar Andersson steal just the cuspid in the 1920's? The ship that was supposed to transport Peking man's fossils to the United States was sunk with all the fossils on board. So what was the mode of transportation for Peking man's tooth from the bottom of the China Sea to Sweden? Could someone have taken a tooth from any collection of teeth and made it look like the tooth from Peking man? I would bet that this is the case far more than I would bet the tooth travelled all the way across Asia, then ground transported itself to the Museum of Evolution at Uppsala University. When I was in dental school, before we worked on actual patients, we spent two years collecting extracted teeth from oral surgeons. We would then set them in metal jaws with plaster, and do our practice drilling. I had jars full of extracted teeth, and there were many cuspids that could easily be made to look just like this one. They came in all shapes and sizes. The story that all of the fossils of Peking man vanished during World War II, then one of the teeth made its way to the very obscure Museum of Evolution at Uppsala University where it sat "untouched since it was dug up in the 1920's" defies credulity. Once a tooth is extracted, if there are no tissue tags or bone remnants, there is no possible way to determine who its original owner was. This is the case with Peking man's cuspid. Of course the overriding issue isn't whether this tooth was a cuspid from Peking man, but how could an evo- illusionist fool so many people into thinking an entire hominid could be made to appear from this and a few other teeth and a few bones that supposedly accompanied Peking man's teeth. Even more stunning is the number of scientists that are fooled by, and support this illusion. Peking man, you are truly an incredible guy. The Peking man illusion has become well accepted throughout the world as one of the most prominent of all hominids.

A museum was constructed at the Zhoukoudian Site in China, which is visited, by tens of thousands of people every year. The museum itself is an incredible illusion. Of course evo-artists turned the few ape teeth and bones found at Zhoukoudian into the intelligent-looking pre-human statue in Figure 3-18; and into a beautiful museum dedicated to those few teeth and bones. See what happens when someone finds a few broken teeth and ape bones in the dirt? As the visitors first approach the museum, they're met with the statue, which is the first part of this well thought out illusion. It will convince visitors before they even enter the museum that this is a hallowed place where pre-humans were found. In their minds, every bone they view inside will be that of the persons represented by the statue; a 700,000 year old hominid. They will be sure of it. The statue has prepared their minds, just like a magician distracts and prepares the minds of his audience. The visitors will walk inside and be certain that those few bones and teeth they see are really from a being represented by the statue, and that they are really the bones of an early human precursor. They will be dazzled. Nearly every visitor will be fooled by the illusion. I have no doubt I would have been. [31-35]

Chapter 4

How to Make a Pre-Human Out of Some Old Ape Bones

Cellophane flowers of yellow and green, towering over your head, look for the girl with the Sun in her eyes, and she's gone...- Lyrics from *Lucy in the Sky With Diamonds*

Earlier in this book I discussed the Scopes Trial, which you may recall was an infamous American legal case in 1925 in which, a substitute high school teacher, John Scopes, was accused of violating Tennessee's Butler Act. The Butler Act made it unlawful to teach human evolution in any state-funded school. The trial was not so much to determine if Scopes was guilty or innocent of teaching evolution. It turned out to be a test case used to force the teaching of evolution in the science classes of public schools, and a challenge for the Butler Act. With the support of evo-illusionists and their followers, Scopes deliberately incriminated himself so that the case could go forward and result in a yay or nay verdict for evolution. Even though evo-illusionists put up a massive defense for John Scopes, he was found guilty. He received a small tongue in cheek fine, which was later rescinded. The Scopes trial was a catalyst, which started a revolution in modern-day science classes and the beginning of universal acceptance of the teaching of evolution in public schools. Scopes makes an interesting divide in the history of human evolution. At the time of the trial, the evidence for human evolution was made up of the hominids discussed in the last three chapters. In other words, it was pathetic. To review, the evidence at the start of the Scopes trial was composed of:

1. 1853: Neanderthal, who was a human, and too young to be a human precursor.

2. 1868: Cro-Magnon man, who was a human.

3. 1891: Java man, composed of a single skullcap, a tooth, and a thighbone; found on the Indonesian island of Java. A second skullcap had not yet been found as of the Scopes trial.

4. 1907: Heidelberg man composed of only a human jaw unearthed in Germany.

5. 1912: Piltdown Prank-man that had not yet been exposed as a prank.

6. 1921: Peking man, composed of four teeth found near Peking, that were metamorphosed into a 700,000-year-old hominid A skullcap and jaw had not yet been found as of the Scopes trial.

7. 1922: Nebraska Tooth, a single "hominid" tooth found in Nebraska that was metamorphosed into the First *Ape Man of the Western World*. Embarrassingly exposed as a pig's tooth just before the trial.

8. 1924: Australopithecus africanus, a partial skull found in Africa, just before the trial. The news of this find didn't make it to the United States in time to have any effect on the trial.

Just think, evo-illusionists and their supporters went to trial with only these few items as evidence for human evolution. It's astounding to consider since there had to be millions of hominid individuals that had populated the planet Earth if evolution were truly valid. Their bones or fossils should be plentiful. The Scopes defense team for the trial must have been panicked. Going into a trial with evidence as pathetic as this had to be disconcerting. Losing Nebraska Tooth-man just before the trial was catastrophic. But the defense went forward anyway. They had one of the best and most famous lawyers in the country, Clarence Darrow, as the defender of evolution and John Scopes. Darrow's reputation and speaking ability went a long way in making up for the lack of evidence for human evolution. Are the more recent fossil finds, the fossils found after the Scopes trial, more effective evidence in proving the evolution of mankind than were the pre-Scopes fossils? We shall soon see.[1,2]

Modern science is incredible in so many ways. Modeling science developed after the Scopes trial made life easier for evo-illusionists. They gained the ability to take the skull of a deceased person, and scientifically refashion the facial structures, and make a living likeness. This technique is a tremendous tool that was developed by police crime labs. It's used when a decayed body or skeleton of a murder victim is found that cannot be identified. Expert artists, sculptors, and anatomists make models of the crime victim's living appearance. Using this same technique, evo-illusionists now have not only clay sculptures, paintings, and drawings made of new fossil finds, but also very real looking models. We now can go to a museum and stand with the real looking hominids that evo-illusionists say are our distant ancestors.

When my wife and I traveled to Stockholm a few years ago, we visited the Vasa Museum, a museum that was dedicated to a seventeenth-century Swedish warship, the *Vasa*, that foundered and sank after sailing only three-quarters of a mile into Stockholm Harbor. Construction on the ship was begun in 1626. The chief engineer designed the entire project in his head. Scandinavian ships were not constructed from drawings then. Instead, shipwrights were given the ship's overall dimensions and requirements. They then conceptually prepared plans based on their experience, common sense, and intelligence, without drawings.

They directed the shipbuilders much like a movie director directs actors in a movie. The King of Sweden, Gustavus Adolphus, demanded there be an additional deck and row of cannon, one more than what the shipwrights designed. The king was warned that if the additional deck and cannon were added, the ship would be top-heavy, and rollover. Of course the king disagreed, and won the argument. The extra deck and cannon were added. A short time after launch the ship rolled over and sank, as predicted by the shipwrights. About 30 men, women, and children died in the catastrophe. The ship lay lost at the bottom of Stockholm harbor for over three centuries. It was located and raised in the 1960's. It was restored and moved to the Vasa Museum in Stockholm in 1988. The skeletons of at least 15 of the victims were found when the ship was brought up. Because fresh water from the nearby mountains constantly flowed into Stockholm Harbor, the wood of the ship and the skeletons of the crew remained fairly intact over those hundreds of years. A post-mortem facial reconstruction lab recreated lifelike models of the victims. (Figure 4-1) It was so

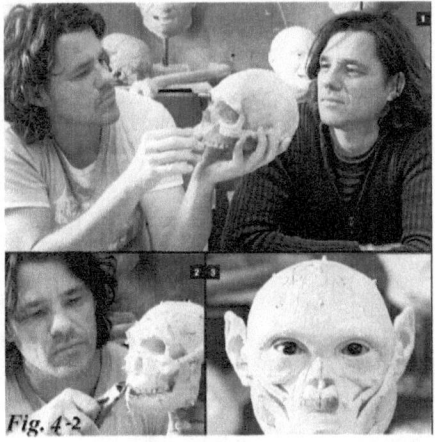

Fig. 4-2

fascinating to be able to see models of the crew of this four-hundred-year-old vessel. It was a bit like taking a trip on a time machine back to the 16th

Century. The models were so well constructed, it almost seemed as if we could talk to them. [3]

Models like those of the Vasa crew are made from supposed hominid bones. Just think, we humans now have the ability to reconstruct the countenance of pre-humans, if they truly existed, that lived hundreds of thousands or even millions of years ago. What could be more fascinating? In the book, *Evolution: The Human Story*,[4] written by Dr. Alice Roberts, photos of models made by *Kennis & Kennis Reconstructions*, an anthropological reconstruction lab, (Figure 4-2) were utilized to bring hominid skulls back to life. On the Kennis & Kennis website they say, "...we only create animals and humans that really existed, and they have to be scientifically accurate." Kennis & Kennis Reconstructions is respected throughout the world of anthropology for their beautiful recreations and paintings.[5] They've done numerous projects for National Geographic, who had this to say:

Fig. 4-3

> *Most paleo-reconstructions of extinct hominids feel stiff and lifeless... but not theirs. The emotion and expression of their Figures is unique. As examples, see their "Lucy's baby"....(Figure 4-3) The playful 3-year-old Australopithecus afarensis was our cover in November, 2006.* [6]

Kennis & Kennis do a beautiful job of creating models of ancient animals and hominids. The models look wonderfully lifelike; so kudos to them for that. But, being the skeptic that I am, I thought it would be interesting to give their models a good scientific analysis. Knowing they had to support the illusions of evo-illusionists, my bet was that I would uncover some of those illusions. There certainly is an incentive for Kennis & Kennis, and all evo-illusion model makers, to bend their reconstructions toward proving the illusion of the evolution of man. Money talks, and so do the vast number of models, sculptures, and paintings concocted by evo-artists and evo-sculptors. The remainder of this chapter will be dedicated to demonstrating that these models, sculptures, and paintings of essentially all hominids on the tree of life in virtually every museum of natural history in the world are nothing but illusions.

The most human-like hominid, Homo ergaster: *Homo ergaster,* who is considered by some evo-illusionists to be a form of Homo erectus, African version, is credited with being the ancestor of humans with the most human-like characteristics of all hominids. Of course he has a cute name: *Working Man*, because of the rock tools left near his remains. If Homo ergaster truly is in the Homo erectus clan, he joins his cousins Java man in Java, Indonesia, and Peking man in China on the illusionary march to Homo sapiens. Homo

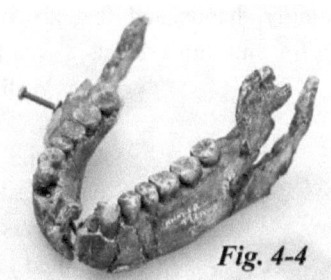

Fig. 4-4

ergaster supposedly roamed the area of Kenya 1.5 million years ago. The first specimen dug up was a lower jaw (Figure 4-4) found in 1971 in East Turkana, Kenya. Again, just think, a whole hominid can be made to magically appear out of just a lower jaw. Evo-illusionists were already experienced at making hominids out of just a few teeth or single jaw as they did with Peking man and Homo heidelbergensis.

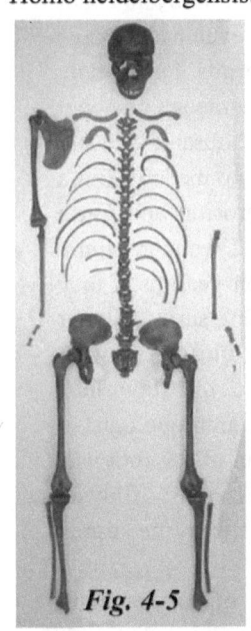

Fig. 4-5

Diggers unearthed other bones of Homo ergaster. The most notable was found thirteen years later, in 1984, in a different area of Kenya. How a jaw found in 1971 could be linked to bones found thirteen years later in a completely different location in Kenya is puzzling. I must keep remembering, this is an illusion. The full Homo ergaster skeleton (Figure 4-5) lacks some arm bones, and other miscellaneous parts. But what's really strange is that both hands and both feet are entirely missing. Hands and feet are a dead giveaway in diagnosing whether a fossil is ape or human. If Homo ergaster were just some old ape bones, he would go on to complete obscurity. If after years of digging, I found a fossil like Homo ergaster, and it had ape hands and feet, I would certainly consider ditching the hands and feet so my fossil would look hominid. A hominid creates fame and fortune; an ape skeleton isn't worth much. What would be dishonest about adding to an already firmly established illusion? In *Evolution, The Human Story*, Dr. Alice Roberts, gives the excuse for no hands and feet: "Finger bones are small, and among the most likely to be lost during fossilization." That excuse doesn't fly, as museums exhibit far smaller and far older fossilized animal bones.

Interestingly, hands and feet are often added to model skeletons of Homo ergaster for museum exhibits. I wonder why. [7-9]

Fig. 4-6

Well, never fear. Evo-illusionists have the answer to every challenge... just about. Anthropologists found a set of footprints in Kenya, across Lake Turkana from where the Homo ergaster fossil was located. Yes, believe it or not, they were credited to Homo ergaster. The footprints (Figure 4-6) were dated at 1.51 million years old. They show big toes that are in line with rather than separated from the other toes. The walker also had high arches, indicative of human-like heel to toe walking. So we can conclude that these 1.51 million-year-old footprints must have come from Homo ergaster. I mean, what else could have caused those footprints in the last million and a half years or so? Fossils of feet and hands of Homo ergaster aren't even needed now since evo-illusionists have Homo ergaster's human-like footprints. For certain. I wonder how they established their age. I mean, how would someone go about estimating how old these footprints really are? Seeing a thousand-year-old footprint would be pretty incredible, especially if you knew who made it. But a 1.51 million-year-old footprint? With all of the rain and erosion and Earth movement that occurred in the last million and a half years? I wonder. Hmm. The only way footprints like these could survive 1.51 million years is if they were made in molten lava. Ouch! Even then they still wouldn't survive. Later lava flows would erase them. But survive they did, in this evo-illusion anyway. Scientific illusionists say volcanic ash may have fallen in the footprints, preserving them. I guess that makes sense. Dan Lieberman, an anthropologist at Harvard University, says "the footprints confirm the evolution of the foot was crucial to becoming human. For one thing, it allowed people to run!" Allowed people to run? I say the footprint couldn't possibly have come from the animal that left Homo ergaster bones; but that's just me. [7-10]

In any case, Homo ergaster's skull is typical of that of an ape, with protruding jaws, large brow ridges, a flat skullcap, and a forehead that isn't vertical enough to house a significant frontal brain lobe. It has all the features of ape skulls. Did Homo ergaster have the skull of an ape, and the feet of modern humans? In this illusion, they sure did. [11-13]

At left in Figure 4-7 is the skull of Homo ergaster constructed from bones found in multiple geographic locations, as is typical of human evo-illusion. At right is the unfinished restoration made by

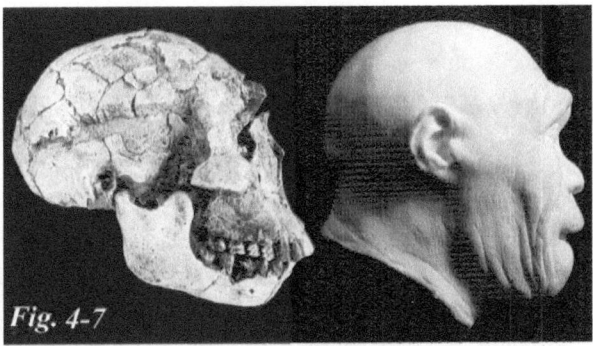

Fig. 4-7

Kennis & Kennis that will be presented in museums and in *Evolution: The Human Story*. It still needs to have skin color and fur or hair added. This near-final step is perfect for analyzing the overall shape and expression of the model. Hair or fur would obscure the shape of the head. The model was given a small human-type ear, not an ape ear. It has small flat jaws that don't follow the shape of the skull; and an abundance of human expression. Mr. Ergaster looks like my favorite uncle, Uncle Chuck. This Uncle Chuck almost looks like he needs a business suit.

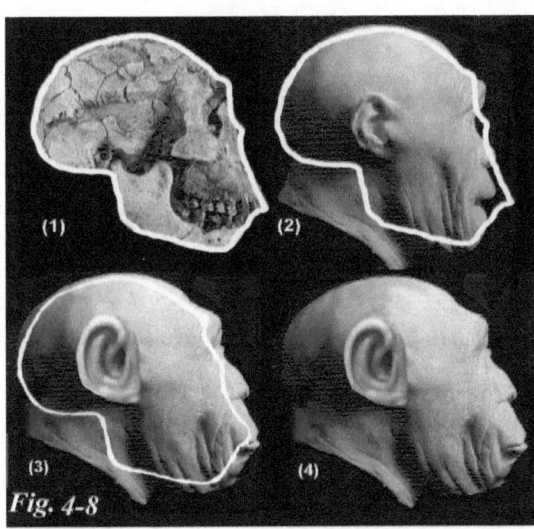

Fig. 4-8

To test the validity of the reconstruction (Figure 4-8) I made an outline of the Homo ergaster skull (1). The outside of the white line is the outline of the skull. I placed the outline over the Kennis & Kennis recreation (2), and, *voila*! The fit is terrible! Of course modeling like this *must* follow the shape of the skull to be at all accurate. The slightest deviation would make Homo ergaster a disaster as far as being an accurate scientific recreation goes. Can you imagine if a police lab made post-mortem models of the remains of murder victims that didn't follow the shape of the skull? No deceased person would ever be identified.

Modeling would be a complete waste. Police lab modelers who did this would be fired. Inaccurate evo-modelers get pats on the back, extra money, and kudos.

You can see that the modelers didn't follow the outline of the skull at all. The jaws of the fossil skull are far more protruded than are the jaws of the Kennis & Kennis model. Exactly what were Kennis & Kennis using for their guide? The skull is also wider front-to-back than the model shows. In making these comparisons, I used the Homo ergaster skull presented in *Evolution: The Human Story.* I also compared numerous other fossil Homo ergaster skulls from other museums to double-check my work. I wanted to make sure there were no Homo ergaster skulls that actually did follow the form of the model made by Kennis & Kennis. The results were all the same. The jaws of each skull I found and tested from other sources were far more prognathic than is the skull pictured in *Evolution: The Human Story.* In (3) of Figure 4-8 I reconstructed the Kennis & Kennis model so it does fit the outline of the skull. In (4) you can see the new model, without the outline. I was very conservative in changing the jaws. I used lateral head x-rays of normal humans to determine how far the lip protrudes from the teeth. To be conservative, I used about 1/3 less than the dimension I measured.

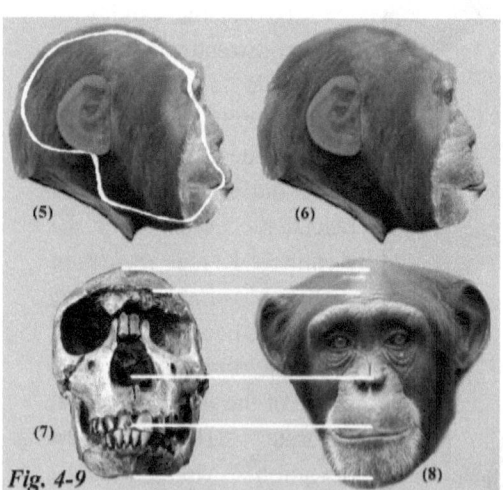

Fig. 4-9

In Figure 4-9 (5) I added ape skin color, fur, and ape eyes. I gave Homo ergaster larger ears that are typical of chimps, since humans are closest to them on evolution's phylogenetic tree. I refined its nose and the width of its head, also to fit the fossil outline. I constructed my version of modeling of the Homo ergaster skull using ape characteristics, because, well, it *was* an ape. So what do we have now? Certainly not the cute expressive pre-human Uncle Chuck as reconstructed by Kennis and Kennis Reconstructions. We have an ape (6). Figure 4-9 (7) is the front view of Homo ergaster's skull, and (8) is the face of a proportionately sized chimpanzee. Notice how the proportions are nearly identical. The chimp actually has a larger forehead than does Mr. ergaster. The chimp would be a better toolmaker for sure.

Why isn't the possibility that Homo ergaster was nothing but an ape at least given as an option by the author, Dr. Alice Roberts, Kennis and Kennis, and all evo-illusionists? Is the obvious fact that the Kennis & Kennis model doesn't fit the fossil skull simply an isolated error? Was this purely an accident, or was it purposeful? Were there discussions about *not* following the shape of the skull so the resulting model will be more human-like?

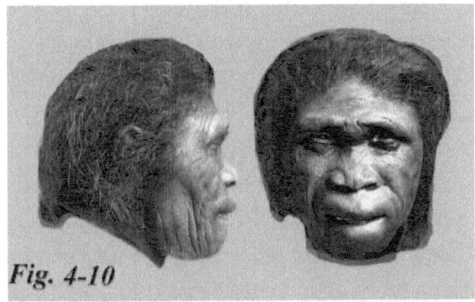

Figure 4-10 is the finished Kennis and Kennis model of Homo ergaster as displayed in *Evolution: The Human Story*. The

Fig. 4-10

hair and skin color were added. Notice the human-like thoughtful expression, the human eyes and hair, and the flat jaws. The Kennis & Kennis reconstruction of Homo ergaster resulted after months of work from the same skull I utilized to make my reconstruction. The expression on the Kennis & Kennis model looks a bit like ex-president Bill Clinton, with that lip bite. Maybe Clinton was the actual model used to make this hominid. If the audience used only the hominid restorations in *Evolution: The Human Story* as criteria, no doubt they would think Homo ergaster is pre-human. In fact every hominid restoration in *Evolution: The Human Story* is given the same treatment. They all look like thoughtful humans. There is no doubt this guy was all ready to go out and make some tools, and maybe even get some computer training. [10-14]

What's really interesting and actually kind of fun is how easy it is to perform the illusion of turning an ape into a hominid by only changing ape eyes to human eyes, as was done on the model of Homo ergaster in Figure 4-10. It really doesn't take much effort, but the results are astonishing. At left in Figure 4-11 is

Fig. 4-11

the face of an ape. Notice, the ape eyes don't show sclera. (the whites of the eyes) By simply giving the ape less ovoid human-type eyes, more white, and a smaller iris, (right) a proximate human face can be made to appear from an ape face. Notice that human eyes show almost 50% sclera. The change is very subtle, but the result is almost magical. Which one would you say is staring, which one is thinking? Which of the two would you say would be the more intelligent? Which might be most likely to make tools, or play pool or chess with you? Which is a recent human ancestor? Of course the ape-person on the right. You will see this technique used by evo-artists on virtually all hominid models and artwork I discuss in this book. Using expressive human eyes alone on hominid models can give the illusion that every fossil find is about ready to evolve into a human. I wonder what survival advantage, or what kind of mutations occurred that changed the large irises of ape eyes to the smaller irises of humans. What mutations greatly increased the display of sclera? I bet any evo-illusionist could answer that question, but I just can't. Now that you know how modern human evo-illusions are conjured, let's move on to the most notable post-Scopes trial fossil finds, and the resulting hominids, and see if evo-illusionists improved their techniques.

It's Spain's Turn: The next on my list of post-Scopes-trial potential human precursors, according to evo-illusionists, is the amazing, and very fun for me, *Homo antecessor*. Homo antecessor is an extinct pre-human species that supposedly lived in what is now Spain from 1.2 million to 800,000 years ago. Again, this is very interesting, since mankind didn't migrate out of Africa until 60,000 years ago. Homo antecessor was discovered in 1994 in

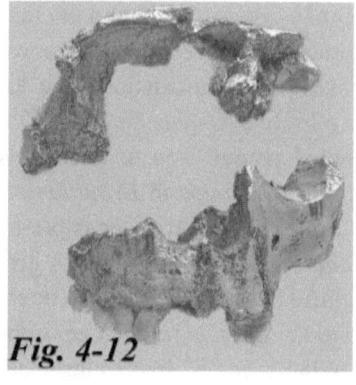

Fig. 4-12

Atapuerca by paleoanthropologist Eudald Carbonell his associates. Fragments of six other Homo antecessors, and hundreds of stone tools and hundreds of animal bones were also unearthed at the site. The best-preserved fossil is a maxilla (upper jaw) and skullcap that belonged to a ten-year-old individual pictured in Figure 4-12. Yes, I said these were the "the best-preserved" out of seven individual fossils. I would ask, why would seven dead people have hundreds of tools? Did these people bury their tools with their dead? Why are hundreds of primitive stone tools so frequently unearthed with dead human precursors? Even more interesting, why are the remains of hominids always just skullcaps buried with their tools and not much more? Where oh where are all of those other bones? Very puzzling.

Homo antecessor's flat skullcap, and complete lack of forehead means he is no different than any ape with a flat skullcap and no forehead. This means Homo antecessor could not invent and manufacture the stone tools it was purportedly found with. This can be concluded with a 100 percent certainty. Which means that the stone tools found around the Homo antecessor fossil were placed there by intelligent humans, which gives the illusion that Homo antecessor made stone tools. Or they aren't stone tools at all, but just some broken rocks that have been carefully selected by intelligent humans, then touted as being stone tools. Remember, stone tools can be put in a bag, and moved or placed wherever the owner of the stone tools wants them. There is no possible way to know if the stone tools were made by the individual that died and left a few bone segments near where the erstwhile stone tools were found.[15]

Evo-illusionists consider Homo antecessor to be incredibly important. *Scientific American* magazine says Homo antecessor evolved into Homo heidelbergensis, which then split, and evolved into Neanderthals and humans. Isn't it amazing that a respected scientific journal could come to that amazing conclusion when there is no possible way to prove anything about Homo antecessor other than it kinda looks like an ape skull fragment?

In 2007, thirteen years after the first Homo antecessor fossil was recovered, Spanish researchers working in the mountains over ten miles away from the first find, announced that they had recovered a molar tooth dated to 1.2 to–1.1 million years ago. The tooth was described as "well worn" and from an individual between 20 and 25 years of age. Of course the tooth was immediately assigned to Homo antecessor. Why wouldn't it be? Any additional teeth or bones found in just about any nearby location, well, within thousands of miles, will be added to the main Homo antecessor illusion, which will make Homo antecessor more believable as a precursor of modern man.

Of course I have questions. Why would paleoanthropologists dig ten miles away from the original find? Did it take an intense amount of digging to find… a tooth? What would make someone who happened to find the tooth in the dirt ten miles away from the site of Homo antecessor's skull and jawbone actually think it came from Homo antecessor? Again, if I happened to be digging a hole somewhere for some reason, and I came across a tooth, the last thing I would think is that it came from a million-year-old human precursor. I would probably toss it in my sock drawer. Maybe the guy who found this tooth was related to the guy that found Nebraska Tooth. But evo-illusionists will go to all lengths, and say anything to conjure sub-illusions needed to prove Homo antecessor evolved into a human. Anytime any evo-illusionist finds or is notified about a found tooth or bony body part, be it pig, ape, or human, it will immediately be transformed into a full human precursor hundreds of thousands of years old. Fame and fortune will soon follow. Additional findings announced in 2008 included the discovery of a small mandible (lower jaw) fragment, and "stone flakes". Stone flakes? Does that mean stone tools? The only known fossils of Homo antecessor are from the two sites in the Sierra de Atapuerca region of northern Spain where the original fossils were found.

If you take a good look at the Homo antecessor skullcap (Figure 4-12) and upper jaw pictured on the previous page, you may note a couple of interesting things. The two pieces don't fit together, as the parts that would show fit are so obviously missing. This should cause a good scientist to wonder if they truly do come from the same individual or species. Unfortunately there is no method of determining if they do, so of course the question is ignored. But, is it possible the maxilla (upper jaw) is from a human and the skullcap is from an ape? Would anyone plant an ape skullcap there to enhance the find of the upper jaw? What happened to the missing parts of that skull? I ask again, as I did with all earlier hominid finds, why did only a couple parts survive for 800,000 years, whilst the rest of the skull and skeleton vanished? One would think if 80 percent of the skull dissolved over time, why didn't the entire skull also dissolve? Instead of wondering if these boney parts came from different individuals, they are inexorably tied together into one individual by evo-illusionists. Remember, this is an illusion, and all is fair in love and illusions.

Fig. 4-13

Homo antecessor's evo-illusionists quickly turned its bone fragments over to the true heroes human evo-illusion: commissioned evo-artists and evo-sculptors. The evo-artists and evo-sculptors took the bones and tooth and made them appear to be full-living proto-humans, with human-type skin, beards, head hair that looks human, and facial expressions that look human as well. Actually

I wonder if the bones were used as models at all. Actually the artwork could have been ordered on the phone. Check out some great examples in the rogue's gallery of illusory Homo antecessors in Figure 4-13. Yes, those are all Homo antecessors created by different evo-artists and evo-sculptors. Homo antecessors are never painted as if they were apes, or with ape characteristics, which is what they were. Homo antecessor's two skull fragments are made to magically appear human. The evo-painters and evo-sculptors are commissioned, so they paint and sculpt as they're instructed. They make it ever more difficult for anyone to doubt or challenge any hominid. The illusory paintings and photos are placed in textbooks, museums, encyclopedias, television documentaries, and lectures. The illusion is a lock. It cannot be challenged or removed. The few bones of Homo antecessor evolved into a pre-human being in only a few short years, not hundreds of thousands of years, thanks to the illusions conjured by evo-illusionists and their dedicated assistants, the artists and sculptors.[12-14]

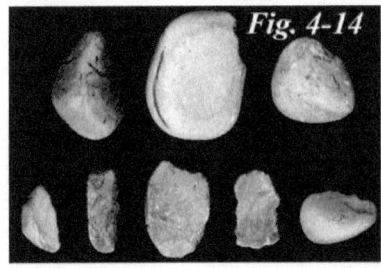
Fig. 4-14

In 2005, to add to the tale of Homo antecessor, flint tools (Figure 4-14) were found in the cliffs at Pakefield near Lowestoft in Suffolk, **England**, over one thousand miles away from Atapuerca where Homo antecessor was found. Do these look like intelligently constructed stone tools? Of course these tools had to be assigned to the Spaniard Homo antecessor. Why not? There weren't any British hominids to assign them to. The closest one is Homo antecessor. Adding to the illusion of Homo antecessor certainly helps in getting audiences to fall for human evo-illusion. Evo-illusionists say the finding of these tools suggests that hominids roamed England 700,000 years ago. So here are Homo antecessor tools with no nearby fossils. Strange. Evo-illusionists say the excavated hewn quartzite tools were used as choppers, to cut meat and wood. One site said,

These tools represent the first instances of technological innovation in human history, wherein our ancestors first began to enhance their biological abilities with the manufacture of stone tools. This speaks to an important milestone in the evolution of our ancestors.[15-17]

If you were walking along a path, and found those stones, would you pick them up and run to the nearest Museum of Natural History, sure that you had found the stone tools of an ancient Hominid? Of course these stone tools lead to the invention of the space shuttle and computer! See how evo-illusionists conjure their illusions? A couple of segments of the skull of a single individual were found in Spain. Some rocks were found in England. The rocks that were unattached to any fossils were conjured into tools made by Homo antecessor, the fossil found in Spain. Remember, this all occurred before the advent of 747's. How did Homo antecessors in Spain travel all the way to the UK? Evo-illusionists fool their audiences into believing the tools were made by Homo antecessor. Does anyone in any audience question? It's very rare that they do. Once an evo-illusion like this is initially accepted by the audience, the evo-illusionists can say or do just about anything they want. The audience will believe. The site went on to note:

The tools were either intimately linked to, or the progenitor of human intelligence, and possibly the cause of the increase in human brain size.

The opportunity for evo-illusionists to exacerbate the value of a couple of bones and a bunch of broken rocks found a thousand miles away has endless possibilities. So the making of the tools can cause the brain to enlarge? This is just another sub-illusion to add to the illusion. What they're saying is the thinking required to make tools causes the brain to enlarge over generations. The illusion doesn't end with tools, teeth, and bones. Use of an organ does not cause inherited increase in size of the organ in the offspring. Muscle builders do not have children with larger muscles than they normally would have. This is an accepted biological law that is frequently ignored by evo-illusionists. It sure is ignored in this case.

The story of Homo antecessor gets even more amazing. In May of 2013, sets of fossilized footprints were discovered in an estuary at Happisburgh on the coast of Norfolk, England. They were credited to be from Homo antecessor dating to more than 800,000 years ago. Again, who else could they be credited to? The UK has no hominids. So the footprints that were found in Norfolk were made by the same people who made the tools in Suffolk, 60 miles away from Norfolk. Both the tools and footprints were made by Homo antecessor who lived in Spain, over one thousand miles away. This certainly makes sense... doesn't it? Do any evo-illusionists wonder how Homo antecessor was able to travel such insane distances? Do they wonder how they crossed the English Channel to get England? Or why would they do such an absurd thing? I think not. These are not important problems for evo-illusionists, since the whole story is just an illusion. National Geographic, certainly a very reputable scientific periodical, in an article by Jane J. Lee, February 7, 2014, gave the huge footprint news this headline: ***Oldest Human Footprints Found Outside of Africa***

The footprint story basically appeared in many periodicals and journals throughout the world. This was a very important find. It was reported with big stories on CBS news, USA Today, the Huffington Post, the Daily Mail, the BBC, the Smithsonian, and dozens of other respected news sources. Of course all of the articles from these respected news organizations were penned without the slightest bit of skepticism.

If modern evo-illusionists said these footprints were from Homo antecessor, well that's it. Why question? Why even think about it? NatGeo went on to say,

As a group of ancient humans walked across a muddy beach in England nearly a million years ago, little did they know that one day, their footsteps would thrill modern discoverers.

Fig. 4-15

Yes, the footprints thrilled modern discoverers! My question is, of course, why would anyone who spotted these footprints (Figure 4-15) think they came from an 800,000 year old Spanish hominid? If I walked down the beach and saw these footprints in the mud, would I be thrilled? I don't think so. I wouldn't think a thing of them, nor would any normal person. I would think they came from a drunk sailor in Her Majesty's Service. If you were taking a stroll along the beach and found these footprints would you rush off to the British Museum of Natural History to sound the alarm?

Well, here is the true story of the find. The finders were actually two scientists, Martin Bates, a professor of history and anthropology, and Nicholas Ashton, who is a curator at the British Museum in London. "We found them by pure chance in May last year," wrote Ashton. When Ashton and Bates were taking a beach stroll, Bates, noticed some hollowed-out holes in hardened sediments, located at the base of a cliff. Bates thought they looked like footprints, so the researchers decided to investigate. Of course they immediately wrote a paper on the astounding find, and published their findings. "We knew the sediments at Happisburgh were over 800,000 years old," says

Ashton. So if the hollows turned out to be footprints, they would be older than anything outside of the cradle of humanity, Africa.

One news periodical, The Daily Tech, was also thrilled with the discovery. They reported:

He (Bates)... *made a fascinating discovery last year that is shedding new light on the path of one of mankind's potential ancestors. Thanks to the follow-up work of a **team of top European museum scholars and university researchers**, these new findings cast light on the footsteps of the hominid nearly a million years ago on the coast of England.* –

So a *team of top European scholars and researchers* was required to solve the mystery of the footprints. I love that word "team". And they were a "top team." It makes everything sound so important. A top team of evo-illusionists can certainly make bigger and more believable illusions than can one single evo-illusionist. How can any normal person question a *top team of top European scholars and researchers*?

So, what was the fate of these fascinating footprints? The tides that uncovered the footprints eroded them into oblivion within two weeks. They no longer exist. They lasted for 800,000 years, they were quickly uncovered by tides, and in two weeks they were washed away. Didn't the rapid disappearance make *the top team* wonder if the

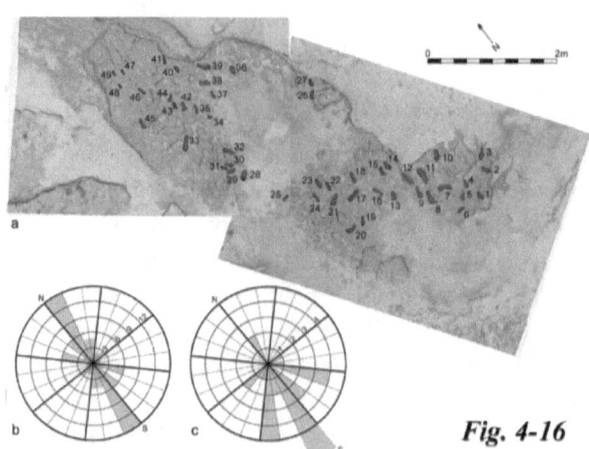

Fig. 4-16

footprints could truly be 800,000 years old? I guess not. What would 800,000 years of rain, wind, and tides do to these muddy footprints? Oh, 'scuse me. I forgot. This is an illusion. In illusions, anything can be claimed and anything can happen. Everyone will agree with the claim, and that's it. Don't ask questions, or you will lose your grant money, your job, and be exfoliated by your fellow evo-illusionists.

Well, anyway, the evo-illusionists had to work fast. Before the footprints vanished the scientists were able to make 3D photogrammetric images and graphic studies of the footprints. Each footprint was carefully mapped, logged, and measured (Figure 4-16). Do people actually get paid for doing this? Just think how valuable these mappings are and will be in the future.

Of course evolution art kicked in to really display and prove the true evolution of Homo antecessor. Figure 4-17 is a scene that was painted showing what the Homo antecessors

Fig.4-17

were doing when they made the footprints. Just think, the original owner of two broken skull segments found in Spain, and his friends, were given a technical name to start a new illusion. Then, in that new illusion, they were made to appear that they made stone tools found in England. They also were made to appear that they left 800,000-year-old footprints, also in England, 60 miles away from where they made the stone tools. Further, they were made to appear that they went to the beach naked, and killed deer. They butchered and ate the deer meat. All of this from those two broken skull bones found in Spain. Once the art is completed, once the *team of top European scholars and researchers* has done its stuff, once he has his name, Homo antecessor is a lock. Homo antecessor cannot be questioned. He is a human precursor for certain.[18-24]

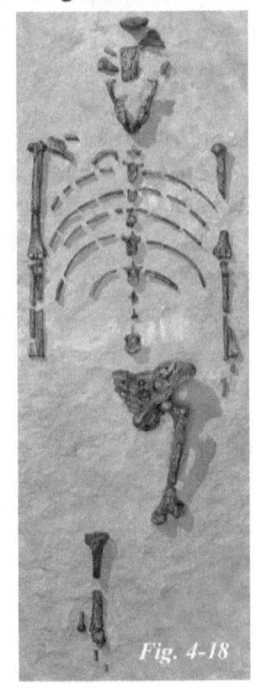

Fig. 4-18

Lucy, another skeleton with no feet or hands: In 1974 one of the most exciting finds in the history of human evolution occurred. I remember reading about the hominid that came to be called *Lucy,* in the Los Angeles Times, and being so fascinated. Lucy got her name because of the popular Beatle song *Lucy in the Sky With Diamonds*, which was a favorite among the team of Lucy diggers. Every time earlier hominids were found, my excitement level increased exponentially. I tried to imagine what it would have been like to take a time machine back millions of years so I could watch Lucy in action. What would these supposed early versions of humanity have been like to actually observe in person? At first thought it would have been fascinating. At second thought, it would have been a complete bore; much like standing too long in front of a chimpanzee cage at the zoo. My interest would have waned quickly. The thing about Lucy that fascinates me is that she had me fooled, just as she fools millions of people across the world. In reality, Lucy was just another on a short list of illusionary human ancestor wannabe's. Amazingly, when I was an evo-devotee, a staunch believer in evolution, the news of each and every new hominid on a very short list had me thrilled.[25]

Lucy's skeleton (Figure 4-18) was unearthed by paleontologist Donald Johanson, who was director of the Cleveland Museum of Natural History, and a graduate student who was working with him. He said he had a "subconscious urge" to dig in a gully in Hadar, Ethiopia. Johanson wrote:

It had been thoroughly checked out at least twice before by other workers, who had found nothing interesting. Nevertheless, conscious of the 'lucky' feeling that had been with me since I woke, I decided to make that small final detour.

I find this story amazing, since Johanson had all of Africa, millions of acres, to use as his digging plot. Why did he choose to dig here, and not in millions of other possible locations? Why did he want to keep digging when it had already been thoroughly dug out twice? Like Java man, and Peking man, Lucy's story is kind of like hitting a hole in one when the golf hole is two miles away. But nonetheless, Johanson's inner voices told him where to dig. Luckily for him, he listened to those voices. After doing some digging, Johanson found the long bone of an arm from what would become the most renowned of hominid finds; one that would make Johanson a famous and wealthy man. Richard Leakey wrote of Johanson's dig:

Johanson had stumbled on a skeleton that was about 40 percent complete, something that is unheard of in human prehistory farther back than about a hundred thousand years. Johanson's hominid had died at least 3 million years ago. But, as additional studies were carried out, it became obvious that this "missing link" was too good to be true.

Too good to be true? Do I detect a bit of jealousy and competition between these world famous paleoanthropologists? Lucy consisted of skull fragments, a partial lower jaw, ribs, an arm bone, a portion of a pelvis, a thighbone, and fragments of shinbones. Johanson also found parts of 34 other adults and 10 infants. But Lucy was the big find.

Johanson of course gave Lucy a complex name: *Australopithecus afarensis*. Who could argue the validity of a very old skeleton once it has been consecrated with both a very cute name, *and* a technical name like that? The names alone boost the illusion that Johanson's find is an ape-person. Remember that if the bones have a technical name *and* cute name when they hit the newspapers, readers will quickly accept the bones as pre-human. That certainly was a selling point for me when I read the story. Johanson concluded that Lucy roamed the Earth 3.5 million years ago. Lucy was declared to be "the earliest definitive evidence of the family Hominidae". Johanson and Lucy were off to the races with this new illusion.

For newly found hominids to gain fame, information about them must be published in scientific journals. Johanson first published his find in *Kirtlandia*, an obscure journal of the Cleveland Museum of Natural History. Johanson and his team of paleoanthropologist allies then bumped Richard Leaky's find, *Australopithecus africanus*, off of the main branch of the family tree of human evolution, and onto a side branch that was a dead end. Can you imagine Leaky's anger when he found out that Johanson arbitrarily bumped his contribution to evolution? Which brings up the question, who does decide where or if the fossils go on the human evolutionary tree? Can one paleoanthropologist with a hominid fossil find simply bump another's hominid off the tree due to his whim, and position of power among other evo-illusionists? What exactly are the criteria for placing a hominid find on the tree in the first place? Does simply painting it on a branch of the tree of life for humans mean for certain it evolved into the next and the next species in a line leading to humans? Again, remember that 99 percent of all species that have ever existed on Earth have gone extinct. What makes primate species so special, and immune to this rule? The 99 percent rule should wipe all found early primate fossils that no longer exist today as modern primates completely off of any tree of life leading to humans. The 99 percent rule means they're nothing but extinct apes. [26,27]

Early on, paleoanthropologists did not think Lucy constructed or used stone tools. However, a 2010 study suggests Lucy and friends were meat eaters. They butchered their hunted carcasses with stone implements. It's pretty hard to imagine a 3 feet 6 inch tall ape making tools, killing other animals, and actually butchering them with tools they made from stones. That's a tough one to swallow. Particularly since virtually all primates are vegetarians except humans. I always love the statement, used so frequently with evo-illusionists: "a study suggests". Anything can "suggest" anything. Anyway, if the study is true, in the world of evo-illusions, hominids used stone tools much earlier than previously suggested: 3.4 million years ago. This is almost a million years earlier than the previous "suggest". The stone tool sub-illusion certainly makes fossil finds more valuable. Which "suggests" there is a competition between fossil finders to conjure tools near a find to make the find exponentially more relevant. Methinks the "study" is just another step in promoting the illusion. Outstanding illusions take many steps. The stone tool sub-illusion is just one of them.[28-30]

Lucy was also credited with being the earliest known bipedal walker. It was "suggested" that she came out of the trees, and foraged for food on the ground by walking upright like modern humans. So Lucy made tools, swung from trees, *and* walked upright. Which, of course, would require an enormous increase in brain capabilities. Remember, of 625 modern primate species, not one single ape has ever been able to form stone tools or walk primarily upright, with the exception of humans. So what is it about Lucy that would bring anyone to think or suggest that she made stone tools and walked upright, other than the need to keep the illusion going?

Bones were supposedly found near where Lucy was discovered that showed scratches indicating Lucy cleaned meat using a hard tool. But the question must be asked, would bones Lucy ate meat from survive for 3.5 million years, when Lucy's skull, her hands, her feet, and most of her skeleton certainly didn't? Is it possible that the scraped bones came from some other source during the last, uh, 3.5 million years? I mean, that's a pretty darn long time. The problem with evo-illusionists' "suggestions" that Lucy was both an upright walker and toolmaker is the fact that her brain size was miniscule. Human brains average about 1400 cubic centimeters in volume. Lucy's brain was about 380 to 430 cubic centimeters. Does this information along with the fact that Lucy had no forehead, a flat cranium, and no significant frontal lobe of her brain suggest anything about her tool-making abilities? It sure does. The true conclusions are obvious, and they should kill the illusion of Lucy. But they don't.

It's so peculiar that out of four chances, two feet plus two hands, like Homo ergaster, Lucy has zero; no feet, and no hands. Is this just another strange coincidence? What I find interesting about Lucy is that a large portion of her skeleton was unearthed, while only a few fragments were unearthed from Java man, Peking man, Homo antecessor, and Homo heidelbergensis. Lucy is

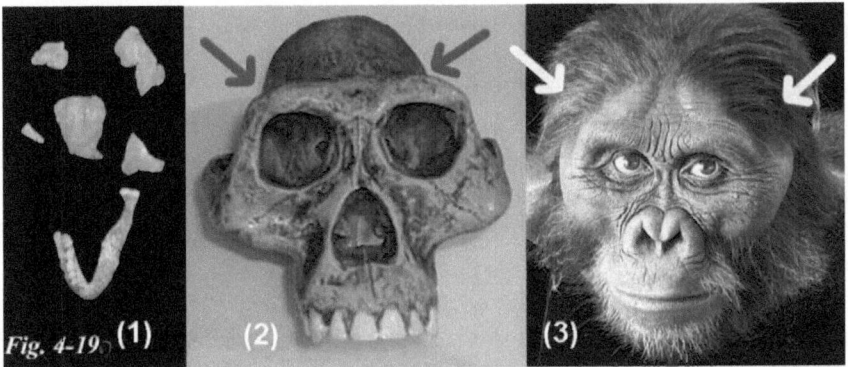

Fig. 4-19 (1)   (2)   (3)

2.8 million years older than Java man, Peking man, Homo antecessor, and Homo heidelbergensis. Almost three million years is a lot of time. Forty percent of Lucy's skeleton was unearthed, whilst only about 3 percent of the skeletons of Java man, Peking man, Homo antecessor, and Homo heidelbergensis were unearthed. Java and Peking men, Homo antecessor, and Homo heidelbergensis are 75 percent younger than Lucy. Wouldn't this mean that most or all of their skeletons should have been found? The rate of decay of fossil bones should be a constant, and an indicator of the age of a fossil. It sure isn't in this case.

If there ever was a presto change-o moment in science, Lucy is it. From the few bones found from Lucy's skull (1 in Figure 4-19) magically appeared the skull in the center (2); with the help of evo-sculptors, of course. Notice the dental arch in (1) is pure ape, not human. From the skull in the center magically appeared that intelligent looking ape-person (3) on the right. Somehow this model doesn't look like a Lucy. How can incredible illusions such as this occur? Of course the answer lies in the only venue where human evolution does occur: evolution art and sculpting. There simply isn't a better example of the illusions conjured by evo-artists and evo-sculptors than that displayed by Lucy. The replica parts of the skull (1) in the montage are housed at the *Museum National d'Histoire Naturelle*, in Paris. As you can see there isn't much. The full skull model in the center (2), which was filled in and constructed by an evo-sculptor, using plaster and lots of imagination, is displayed at the Museum of Man, in San Diego, California. The Smithsonian National Museum of Natural History, utilizing the five chips of bone (1), the imaginary skull in the center (2), and evo-artists and evo-sculptors, now proudly displays the model illusion of what Lucy semi-actually looked like. (3) Yes, folks, again, your tax dollars at work.

Notice that the large and wide cranium and forehead on the model appeared out of nowhere. The cranium on the skull terminates well inside of the eye sockets, (arrows) whilst on the model they're way way outside of the eye sockets. They're so far lateral to the eye sockets, I had the give them a double "way". Notice how the immense brow ridges on the skull have flattened out, and almost vanished into thin air on the model. This is certainly scientific illusion at its best. Notice the eyes on the model show about half iris and half sclera (white), like human eyes. And no one has to wait millions of years to see this happen. Right before your disbelieving eyes, on this very page, evo-illusionists evolve five bone chips and a partial jaw into a complete skull. Then the complete skull is evolved into an intelligent ape-person. Doesn't she look like she's posing for her senior picture? She certainly has human expression and an intelligent, almost human-looking face. I bet she was homecoming queen. This must have been the model that created the "suggestion" that Lucy was a toolmaker and upright walker. It's amazing how evo-artists can make intelligence and expression magically appear out of what was originally just six bone chips.

Hundreds of other bone fragments claimed to be from Australopithecus afarensis were also found at different times in different locations. One was a complete skull (Figure 4-20) of a three-year-old girl. It was unearthed in Salam, Ethiopia, a few miles from where Lucy was found. The remains include this skull, milk teeth, tiny fingers, a torso, a foot, and a kneecap no bigger than a dried pea. This fossil was found to be tens of thousands of years older than Lucy. Even so, she has been named *Lucy's Baby*. It's so strange that her tiny fingers, a kneecap the size of a pea, and a foot survived tens of thousands of years longer than

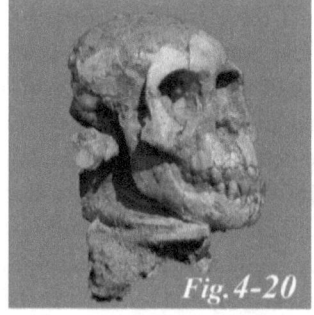

Fig. 4-20

did Lucy's hands and feet. If you compare Baby Lucy's skull with Lucy's skull in Figure 4-19 center, does this skull look like it came from the same species as Lucy? Of course not. But for the sake of the illusion, evo-illusionists will say it did. In a different dig, a separate foot bone was found that has a large arch, which was also credited with belonging to Australopithecus afarensis. It supposedly showed Lucy to be an upright walker. Again, it's interesting to note that Lucy's hands and feet completely vanished from her skeleton. Not one bone survived fossilization, but a foot found in a different location did survive. What happened to the rest of the skeleton that belonged to the foot? It certainly should be present, since the feet erode first. So Lucy has a foot with no skeleton, and a skeleton with no feet.[31]

To me the most astounding part of the Lucy story occurred on national television, on a scientific documentary by NOVA produced by PBS, titled *In Search of Human Origins*. To make things really scientific, Don Johanson, the discoverer of Lucy, was the narrator. On this show he gave the greatest example of evo-illusion, and how it works, that I could possibly imagine. So I want to personally thank him. When I watched this documentary for the first time, I almost thought I was dreaming. This was a real eye-popping, but fun moment. Johanson was discussing Lucy's pelvis. He was working with evo-illusionary anthropologist and "grinder" Owen Lovejoy. Said Johanson,

The knee looked human, but the shape of her hip didn't... Superficially, her hip resembled that of a chimpanzee's, which meant that Lucy couldn't possibly have walked like a modern human. But Lovejoy noticed something odd about the way the bones had been fossilized.

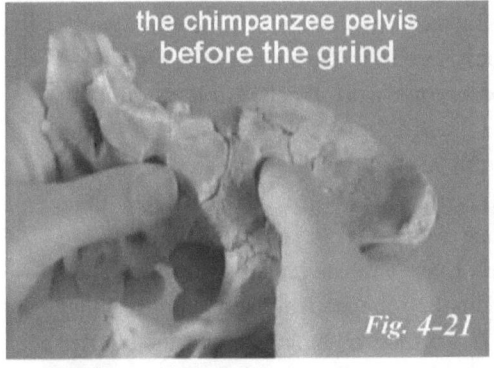
the chimpanzee pelvis
before the grind

Fig. 4-21

Me: Well, maybe Lucy *was* a chimpanzee.

Lovejoy said,

When I put the two parts of the pelvis together that we had, this part of the pelvis has pressed so hard and so completely into this one that it caused it to be broken into a series of individual pieces, which were then fused together in later fossilization.

Me: Take a look at Figure 4-21. Looks like an excellent fit to me. This is the "bad" version of Lucy's pelvis; the one that needed correcting, because it was "like the pelvis of a chimp".

Johanson continues,

After Lucy died, some of her bones lying in the mud must have been crushed or broken, perhaps by animals browsing at the lakeshore... The perfect fit was an illusion that made Lucy's hipbones seem to flair out like a chimp's.

But all was not lost. Lovejoy decided he could restore the pelvis to its "natural shape". He didn't want to tamper with the original, so he made a copy in plaster. He cut the damaged pieces out and put them back together the way they were before Lucy died.

How does Lovejoy know how the pelvis looked before Lucy died, and, that it was crushed into perfect "chimp" position? What are the odds of that happening? Lovejoy took a Dremel tool, and whacked away at the fossil model, and came up with the pelvis

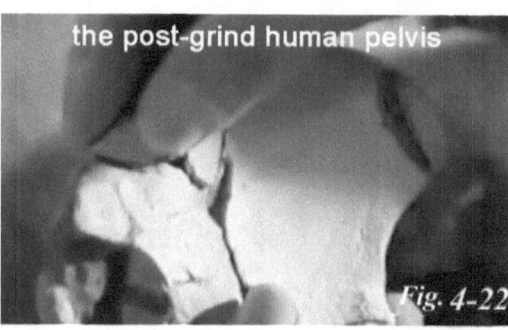

the post-grind human pelvis

Fig. 4-22

pictured in Figure 4-22, which is now a "perfect fit". Compare that with the chimp-like pelvis above. You could read my book through the cracks in the newly ground pelvis. See how evo-illusions work? Evolution doesn't exist in reality. It exists in the illusions of thousands of evo-artists, evo-sculptors, and evo-illusionists like Owen Lovejoy who uses a Dremel tool to quickly evolve a chimp pelvis into a human pelvis. In this case, evolution only takes minutes, not millions of years. Do these guys fool anyone in their audience? Of course they do; they probably fooled nearly everyone who watched this astounding documentary.

There are numerous other characteristics of Lucy that make her nothing but an extinct ape. Knuckle-walking apes have a wrist feature that locks their wrists in place for greater stability when walking. Locked or stiff wrists are amenable to the knuckle walking of apes. Lucy's wrists have this feature. They are not flexible like the wrists of a human. Johanson himself admitted that Lucy's arms had the length of a knuckle-walking chimpanzee, not the shorter arms of a human. Lucy's rib cage is shaped conically, like the rib cage of an ape, not barrel-shaped like those of a human. There other features of Lucy that scream "ape", but these are sufficient. No human has the pelvis, skull, wrists, arm proportions, protruding jaws, or rib cage of Lucy.

Ignoring the fact that Lucy showed all the signs of being ape, which a good illusionist should do, Johanson said:

What we do know is that these creatures were walking like us over three million years ago. And that was a distinct advantage. They could cover long distances, forage for food, and carry it back, perhaps to a faithful mating partner. We believe Lucy's species was the root of the human family tree. She is our earliest ancestor, the missing link between ape and human.

Does Johanson sound any different than the evo-illusionists who were so excited about Piltdown Prank-man and Nebraska Tooth-man? Johanson, the finder himself, was humbly the evo-illusionist that got to declare Lucy was hominid. What could be more scientifically objective? The person who was in the position of making the greatest financial and career gains if Lucy were declared a human ancestor was the decider if Lucy truly was a human ancestor. If enough people believe what he said, and if Lucy stays on the family trees of textbooks and scientific periodicals, Johanson is in. It turns out he is in. On the most up to date version of the human family tree on the Smithsonian

Fig. 4-23

National Museum of Natural History website, there she is, Lucy, in a direct line to Homo sapiens. She's also on the most recent versions of the human family tree in Scientific American. In fact Lucy is prominently situated on the branch

Fig. 4-24

of the family tree that leads to humans on all family trees that I searched out. Lucy's illusion is pervasive. [31]

Johanson went on to paleoanthropological fame. He was on television talk shows and documentaries. He gave numerous lectures to university students and science groups; and of course he wrote books that became best sellers. Notice how his book cover for *Lucy's Child* (Figure 4-23) has a painting by an evo-artist who took the obvious chimpanzee fossil of Lucy, and mystically turned it into a human, with human skin, hair, expression, human posture... Evo-illusions are everywhere in this venue. It matters not if this picture is of Lucy herself, or her descendants. The inference is obvious and overwhelming. Lucy was pre-human, and was the mother of all of humanity. Johanson has become the leading worldwide expert on the evolution of Homo sapiens. Anything he says is now gold; pure unadulterated valid science.[32]

Figure 4-24 is a photo of Johanson and his evo-devotees excitedly standing over a reproduction of Lucy's fossil. Do you think any of them were *not* fooled by the Lucy illusion? Johanson did have moments of rationality and honesty. But they were fleeting, or Lucy wouldn't be on the family tree of humans all over the world. Johanson himself summed up what I have been trying to say. He makes it obvious what the driving force for evolution really is:

There is no such thing as a total lack of bias. I have it; everybody has it. The fossil hunter in the field has it.... In everybody who is looking for hominids, there is a strong urge to learn more about where the human line started. I was working back at around three million... that is very seductive, because you begin to get an idea that that is where Homo did start. You begin straining your eyes to find Homo traits in fossils of that age.... Logical, maybe, but also biased. I was trying to jam evidence of dates into a pattern that would support conclusions about fossils which, on closer inspection, the fossils themselves would not sustain... It is hard for me now to admit how tangled in that thicket I was. But the insidious thing about bias is that it does make one deaf to the cries of other evidence.[32]

I would change the term "bias" to "great need for pats on the back from fellow evo-illusionists, fame, and fortune", all of which were his with the publishing of the Lucy fossil. The need for pats on the back, fame, and fortune create bias, and are the driving force behind all scientific illusions.

Arguments continue regarding whether Lucy was an upright walker, whether her pelvis and other anatomical parts showed bipedal walking, whether she was a toolmaker, whether she could think, hunt and butcher prey with her tools. All of these arguments are just a huge part of the Lucy illusion. It doesn't matter if Lucy was an ape that walked upright. It doesn't matter if Lucy made tools, and butchered her prey with them. It doesn't matter. Why? Because there is absolutely no possible way to prove the infinitely unlikely possibility that Lucy's descendants evolved into human beings. The immense point is, it cannot be proved that any ape-like fossil, no matter how talented it may have been, was from a species that evolved into humans. Period. If Lucy had lots of talents that no ape has today, that in no way proves she morphed into mankind. [33-38]

Homo habilis: *Homo habilis* theoretically roamed Eastern and Southern Africa 2.4 to 1.4 million years ago. The first step in the molding of *Homo habilis* into a human ancestor began in 1959 when two teeth were unearthed at Olduvai Gorge in Tanzania by Louis and Mary Leakey and their hard-working fossil diggers. See? All it takes are a couple of teeth, and an entire hominid is in the making. Remember Peking man? The Olduvai Gorge is thirty miles long and 300 ft. deep. Can you imagine taking a shovel and digging in an area that large, in hopes of finding bones? It would take me about seven minutes for morbid boredom to set in. I certainly must give these fossil diggers a big pat on the back, no matter what their motive is for digging.

The next year, parts of a boy's skeleton were located at the site. Additional fossil bones, such as a lower leg bone and a foot bone from other individuals, continued to be found and added to the original fossil. As is typical, a deformed cranium was found at one location, and jaw and bone fragments at another.

As expected, stone tools were found near the site. Homo habilis is yet another dead guy with lots of tools. The *tools-near-dead-guy* feature is almost a common denominator with hominid finds. As a result of his phenomenal toolmaking ability, Homo habilis was given the cute name *Handy Man*. Again I wonder why tools were made and then left randomly in the dirt near where the toolmakers happened to die. Homo habilis had a brain size of about 640 cubic centimeters, less than half that of humans. With the miniscule size of his brain, and the complete lack of a significant prefrontal brain lobe, Homo habilis cannot have been the toolmaker he was touted to be. Again, it's amazing that there isn't one modern ape species, or any species of any kind for that matter except humans, that can make the very simplest of tools. Why did apes learn to make tools millions of years ago, but not one modern ape has the slightest inkling even what a tool is, nor do they show any interest to make one?

One paper said the cranium size of their specimen gave the Leakys hope they had found a new type of human ancestor. If you unearthed a skull with a 640 cubic centimeter cranium, would you be excited that you had found a human precursor? Or an ape. I would pick ape. After years of digging, what a disappointment it must have really been when the Leakys first saw their "hominid" skull. They probably had to show a lot of fake excitement. So *did* the Leakys find an ape? It really doesn't matter, as Homo habilis has been permanently placed on most human evolutionary timelines including the ones at the Smithsonian and Scientific American. Homo habilis is in! Handy Man has officially been declared to be our ancestor. Some would say that it has been placed on a dead branch of human evolution. But if that's the case, why is it there at all? If it's on a dead branch, it's simply an extinct ape that died near some broken rocks. It has nothing to do with the origin of humans. Homo habilis is on the human timeline because it adds to the illusion of human evolution.

Compare the skull at left with the model from Kennis & Kennis, displayed in *Evolution: The Human Story* in Figure 4-25. Notice the

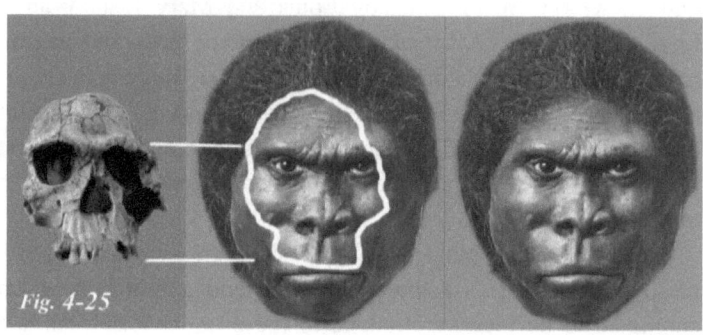

Fig. 4-25

difference between the size of the forehead on the model and on the skull. Ditto the brow ridges. Right behind the brow ridges on the skull are concave spaces, which aren't present on the model. On the skull, the brow ridges are far wider than its tiny cranium; but on the model the temples are far wider than the eye sockets, or what is left of the reduced brow ridges. The cranium is huge.

I drew lines from the tip of the front teeth and the brow ridges on the skull to those on the model so I could proportion the size of the skull to the model. I placed a proportioned outline of the Homo habilis skull over the model (middle image);

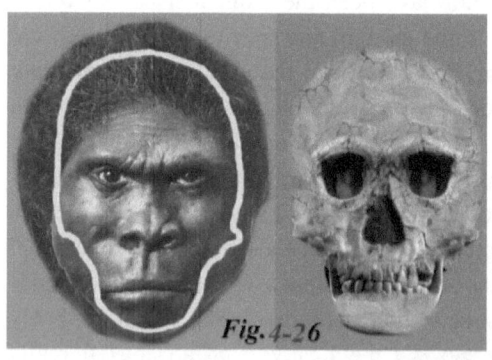

Fig. 4-26

and, well... The model doesn't come close to following the shape of the skull. It's truly astonishing. One wonders what was used as a guide to make the model. I can't imagine. Of course the model gets human eyes with 50% sclera displayed. It has wider human-like lips. The lips of apes are very thin. The model has human hair that would require a haircut at Supercuts on a monthly basis. And it has an angry human expression. The skull of an ape has been mystically transformed into a human precursor.

In Figure 4-26, left image, I made an outline of the skull the Kennis & Kennis model would actually have, since they didn't follow the shape of the Homo habilis skull at all. I then constructed a skull using that outline. I placed the various anatomical parts, like teeth, nose, and eyes, exactly over those on the model. And what was the result? Incredibly, the model head has the shape of a completely human skull (right image in Fig. 4-26). Maybe when they made this model, they accidentally picked up a human skull instead of Homo habilis? Or, is this evo-illusion at work?

Since Homo habilis lived between 2.4 and 1.4 million years ago, this signals another reversal of the time clock of fossil decay. Homo habilis is two to three

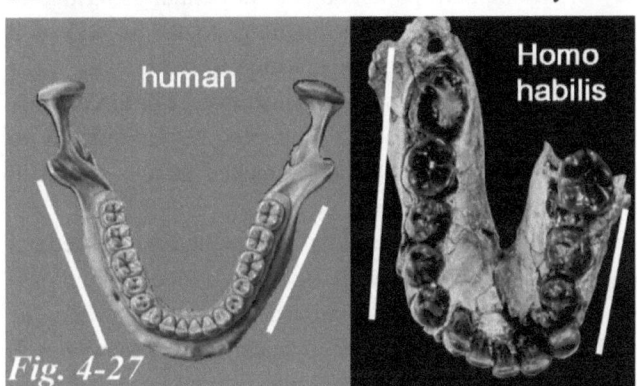
Fig. 4-27

times older than his cousins, Java man, Peking man, Homo antecessor, and Homo heidelbergensis, yet he left almost a full skull whilst they left only fragments. What's helpful about Homo habilis is that he left his nearly complete

mandible (lower jaw). (Figure 4-27) The dental arch is "U" shaped, and not "V" shaped, which indicates Homo habilis was an ape. In 1986 a 1.8-million-year-old partial Homo habilis skeleton was uncovered in Olduvai Gorge. It was made up of 302 fragments of fossilized teeth and bones. Some of the *best* samples are pictured in Figure 4-28. These bones allowed scientists to determine Homo habilis' arm, leg and body proportions. They showed Homo habilis had definite ape-like, not human-like, limb proportions. So the length of Homo habilis'

long bones when compared to his height makes him ape. His high brow ridges, protruded jaws, U-shaped dental arch, miniscule brain size, and lack of a frontal lobe all say Homo habilis was just an ape. Even so, evo-illusionists almost universally accept Homo habilis as a direct human ancestor. The bones that show him to be an ape must push him hundreds of thousands of years further back on the timeline of human evolution. The more ape-like a hominid fossil is, the older it must be. By far the greatest possibility is that Homo habilis was simply an extinct ape; maybe not even extinct. But evo-illusionists cannot even consider this option, nor is it ever

Fig. 4-28

suggested as an option. Instead of being classified as an ape, the Homo habilis illusion has been utilized to christen him to be the first *Homo* ancestor of modern humans. Evo-illusionists promote if a certain specimen of hominid has a bit larger of a braincase than that of another hominid, or if it has smaller brow ridges, these sizes represent the entire population of that species; and that it can then be accurately placed on a timeline of human evolution. This just isn't the case. If any scientist made a major study of any modern populations of any kind, using one or two individuals of that population, they would be a laughing stock. So why is it acceptable for evo-illusionists to do this? The bigger question to me is, why was I so accepting of this practice for so many years? Evo-illusionists would certainly give the excuse that there just aren't many fossil examples of pre-humans to analyze. But then there shouldn't be a theory that is so locked in it is taught to millions of school children all over the world.[39-44]

In reality, each pre-human fossil is an anecdotal representation. Anecdotal evidence isn't valid in any scientific venue except in the world of evolution's illusions, where it's revered. It can fool audiences, and it does. Going by what one or two individuals looked like or acted like is not a valid scientific representation. It takes hundreds or thousands of individuals to have an honest sampling. Displaying one or two skeletons that happen to have died near some broken rocks does not make those one or two individuals toolmakers. If a bow and arrow was found near a gorilla skeleton, could it be concluded that the gorilla made the bow and arrow? Relative brain size, jaw protrusion, and the shape of brow ridges or bones of a couple of individuals cannot be used to measure trends of entire populations. Doing that is just bad science, bad math, *and* bad statistics. But it's good for the illusion of evolution. This is what evo-illusionists have done with their brand of human origins. They have taken very few fossils, painted them into a timeline of human evolution, and declared them to be human ancestors. Which means that placement of all of the supposed hominid species on a timeline that purports to show how modern humans evolved from earlier primates is nothing but an illusion.

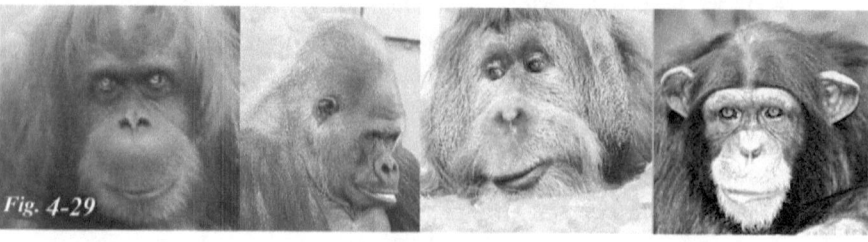

Fig. 4-29

If you take a look at the four apes in Figure 4-29, you can note the differences in brow ridge sizes, cranium sizes, jaw protrusions, and forehead sizes. All of these apes exist today as modern species. If evo-illusionists had unearthed dirty partially destroyed skeletons of these apes, they no doubt would take measurements of their features and place them on an evolutionary timeline. Do any of them look like they could make tools?

Chapter 5

The Story of Hominid Finds Gets More Amazing

When you come to a fork in the road, take it- Yogi Berra

There must have been millions of skeletons left by hominids, if they truly existed, for paleontologists to discover. In reality, hominid finds, or should I say, bones that can be touted as hominid finds, are incredibly rare and an embarrassment to the science that promotes evolution. There are only a few members of the human tree of life that I haven't yet discussed. As you can see from the last three chapters, each hominid has its own fascinating story. There are very basic common denominators among all hominids. They all have skulls with tiny braincases, large brow ridges, extruding jaws, "U" shaped dental arches, and lack vertical foreheads; which means they're apes. They virtually all follow my *Ten Steps to Hominid-ism*. Most have cute names, which make them acceptable to the audience. They also universally have complex names, which confuse the audience so they can't quite understand who they really are, and where they belong on the timeline of human evolution; which is good for human evo-illusion. All hominids have evo-artists and evo- sculptors who greatly exaggerate their models and paintings to make fossil finds look as human as possible. The few remaining hominids I will discuss in this chapter are no exception. They follow my Ten Steps to a tee.

Fig. 5-1   gorilla   P. boisei

The Leaky's find an ape in Tanzania: In 1955, Louis and Mary Leaky, the ultimate human fossil finders, unearthed bones they claimed were hominid, in good old Olduvai Gorge, Tanzania. They gave their bones the complex name *Paranthropus boisei.* Mr. boisei supposedly lived around 2.3 to 1.2 million years ago. He was noted for being a robust chewer because of his large teeth with thick enamel. Figure 5-1 far left is a gorilla skull with a large sagittal bony crest exactly like that of Paranthropus boisei. Sagittal crests make solid attachments for strong muscles of mastication, and are found on other modern apes. Not one human on Earth has that feature. So again, a sagittal crest alone should delineate ape skulls from human skulls.

Paranthropus boisei has all of the other common features of apes, such as large brow ridges, protrusive jaws, an extremely flat and small cranium, a "U" shaped dental arch, and a miniscule forehead that could not house a substantial frontal brain lobe. Paranthropus boisei was an ape. With such a flat face, and huge brow ridges, Paranthropus boisei is a difficult skull for evo-artists and evo-sculptors to evolve into a human, with human emotions and expression. Both reconstructions to the right in Figure 5-1 look pretty angry. Evo-modelers made Mr. Boise's eyes human, but other than that, they had a pretty rough time making him look pre-human, even with the artistic license they routinely utilize. Maybe they should have switched to a human skull like they did when

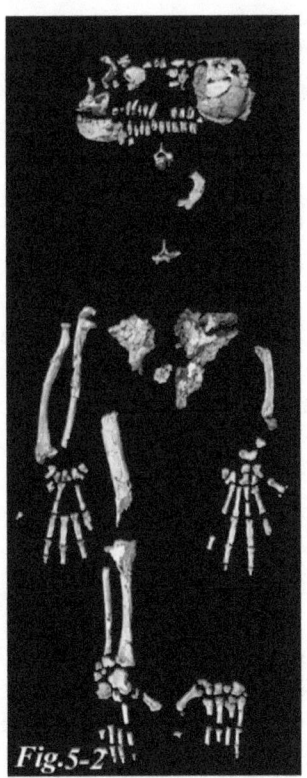

Homo habilis was modeled. There just wasn't a whole lot to work with. He's a tough illusion to conjure, and a pretty grumpy looking illusion at that.[1,2]

Ardi, the common ancestor to all chimps and us: Fossil diggers unearthed a mandible and partial skeleton of an individual, in the Middle Awash area of Ethiopia between 1992 and 1994. Paleoanthropologist Tim White from the University of California, Berkeley, headed the group of diggers. The cranium, jaw, and teeth were quickly determined to be hominid because the bones were given a cute name, *Ardi*, and a complex name: *Ardipithecus ramidus*. Ardi is in a tie with Lucy for the cutest of the cute names of all hominids. Further digs turned up other skeletal parts. Forty percent of an entire fossil skeleton was found. As is typical of evo-illusion, bones unearthed after the original find weren't attached to the original fossil. Notice with Ardi's skeleton in Figure 5-2, none of the joints actually articulate (fit together). This skeleton is no doubt a

Fig.5-2

conglomeration of the bones of many individuals, which of course removes the credibility of Ardi being a pre-human. Also notice that the big toes are separated from the other toes, and they protrude out at 65 degrees. Ardi's feet were ape. Ardi lived 4.4 million years ago, and has been bestowed the honor of being the last common ancestor of humans and chimps. You see, here is another example of how a fossil's mere existence will enhance the illusion that Ardi is a human ancestor. How can it be proved that Ardi truly is our ancestor? The answer is, there is no possible way to determine that Ardi's descendants evolved into humans. It's kind of like a given in geometry. Find a primate fossil; it's a given that it's a human ancestor. That's it. Instead of being just an old ape fossil, Ardi becomes a part of the immense and expanding illusion of human evolution. Any old primate (ape) bones that are dug up become human precursors just about the minute the Sun hits them. Notice how Ardi's artistic recreation in Figure 5-3 is given human eyes, and a completely upright stance. Remember, she has ape hands and ape feet used for grasping, not upright walking

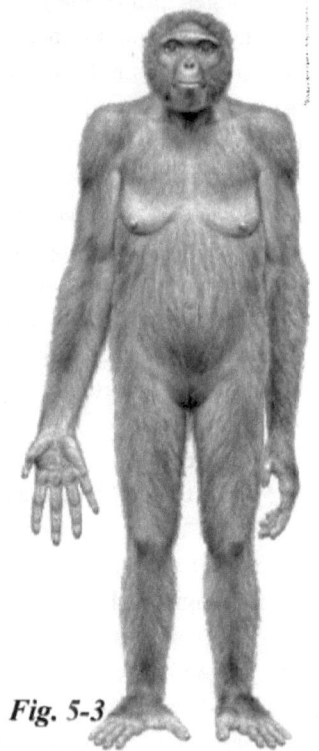

Ardi almost makes me want to go to Africa and dig dig dig for fossils, fame, and fortune. If I found any old ape bones, or any pieces of a skeleton, I can give them a cute name. How about Poofie? I can give them a *Fig. 5-3* complex name, oh, like maybe *Australopithecus pooferoundious?* I can have my wife, an excellent artist, do a few paintings of Poofie, and get them placed on a timeline of human evolution. It just doesn't seem all that tough. All I have to do is dig up only one of the billions of hominid/ape bones that must be buried beneath the soil in Africa. Just a jawbone, or even a partial jawbone, or a broken piece of skull will do. Hey, maybe even a couple of teeth. I could be on late-night talk shows, and get rousing applause. I could write a book; a best-seller for certain. For a title, how about *A Mother Named Poofie?* I could get thousands of pats on the back. I could lecture at major universities. Hmmm. Ya know, I would almost go to

Africa and give it a go. Except I think I'd rather play tennis right now; so maybe next year.

Ardi has been strategically placed on practically every timeline of the tree of life for Homo sapiens all over the world. By giving Ardi both a cute name, and a complex name, and placing it on the human tree of life, she is well entrenched in the illusion of human evolution.

All you have to do is look at Ardi's characteristics, and you can easily discern that Ardi is just another ape. As I noted, Ardi's big toe is very divergent and spread 65⁰ from the other toes, just like that of any other ape. Its thumbs don't reach the first joint of its index fingers, which makes them ape. More

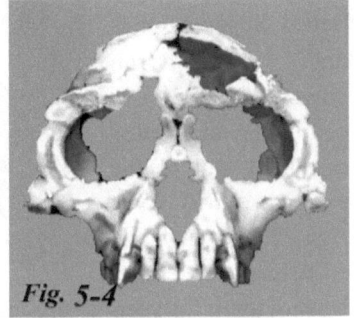

Fig. 5-4

importantly, Ardi's skull (Figure 5-4) shows no forehead that could house a large frontal lobe. It has large brow ridges, prognathic jaws, and a very small flat cranium, all characteristics of apes. Even Ardi's arms are very long, like an ape's. Unfortunately no tools were found with Ardi for me to challenge, and have fun with. But since Ardi's importance has already been established because Ardi was bestowed the title *common ancestor of chimps and humans*, she doesn't need tools. Ardi's discoverer has gained fame and back pats without the tools.

Ardi's pelvis, which was reconstructed from a crushed specimen, is said to show adaptations that combine tree-climbing and bipedal activity. Evo-illusionists have already demonstrated their reliability at handling the pelvis of fossils. Remember how NOVA handled Lucy's pelvis? A little grinding here, a little grinding there, and evolving a chimp pelvis into a human pelvis is an easy task. Evo-illusionists want so badly for Ardi to be bipedal. But this wouldn't really be a good thing for evolution, as later hominid specimens had not yet evolved upright walking. If Ardi were an upright walker, it would spoil the illusion and the timeline. You can't have ancestors that were upright walkers, then their descendants were knuckle walkers; then modern humans becoming upright walkers again. Paleontologists should hope Ardi was a knuckle walker for the sake of the illusion. Let's hope knuckle-walking doesn't evolve back into the human population. That would not be fun.[3-5]

The Gift That Keeps On Giving: In 2013 an Arizona State University student, Chalachew Seyoum, was walking along an archeological dig in Ethiopia when he spotted a jaw poking out of the slope. (Figure 5-5) What amazing luck! Arizona State University scientists

Fig. 5-5

had been digging for years at this site, with little results. But the finding of this jawbone made all of those years of digging worthwhile. According to the New York Daily News:

The ASU team was also able to establish that the owner of the jawbone walked on two legs and lived in a dry, arid climate. Researchers are still trying to establish what it ate and whether it used stone tools.

It's amazing how just one little piece of jawbone with six teeth can yield so much information. Paleoanthropologists are certain the genus of the new find is Homo. The New York Times wrote:

Mr. Seyoum, a graduate student in paleoanthropology at Arizona State University, had made a discovery that vaulted evolutionary science over a barren stretch of fossil record between two million and three million years ago.

Fig. 5-6

See how these illusions grow? What made this find so easy to date? How did they know it was 2.8 million years old?

Well, it was found in sediment that was 2.67 million years old; that's how. And if the sediment is 2.67 million years old, that makes the jawbone 2.8 million years old. Right? It's unthinkable that in the last 2.8 million years any jawbone could have been left on this hill, or buried, by any other means. Isn't it? So a whole new human precursor is born out of just a jawbone fragment. There is no doubt this jawbone fragment was a hominid that evolved into another hominid, and then another, and eventually evolved into human beings. Right? I want to thank modern science for creating these illusions that make us, well, at least many of us humans think we evolved from apes. And all it takes is a fragment of a jawbone to accomplish the task. Just think, in this immense area pictured in Figure 5-6, a student from Arizona State University found that one single piece of broken jawbone.

If you were there, where would you get off your camel and start digging? Where would you pick? Digs have been made here for years with no finds. I'm certain that soon the jawbone will be given a cute name. How about *Arizi*? For the complex name; how about *Australopithecus arizonapus*? Evo-artists will take over and turn the jawbone fragment into a full-sized human precursor, and all of this will wind up as models of hominids in museums, and in students textbooks all over the world. Students will soon be taking

Fig. 5-7

tests on *Australopithecus arizonapus*. "Who discovered *Australopithecus arizonapus* in the deserts of Ethiopia in…" [6,7]

Homo sediba, found by a kid: In searching for other examples of the use of evolution art to conjure illusions, I went to my favorite scientific journal, National Geographic. NatGeo is really my favorite nature journal except when it discusses evolution. Then it becomes my *most* favorite book of evo-illusions. I was certain that it would have some great examples of the illusion of using human eyes and expressions on hominid reconstructions to make apes look human; and bingo, it did! A nine-year-old boy named Mathew Berger, with his dog Tau, found a new hominid fossil in a cave in Malapa, South Africa. Matthew is the son of the now famous paleoanthropologist Lee Berger, who is discussed later in this chapter because he is linked to another astounding hominid unearthing. Mathew's new find was bestowed a scientific name:

Australopithecus sediba, which locks it in as a human precursor. He roamed the Earth about 2 million years ago. From a ragged but typical ape skull with no mandible (Figure 5-7), evo-sculptor and modeler John Gurche evolved Australopithecus sediba into the human precursor pictured at right in Figure 5-7. By giving it a nice human smile, human eyes with all of that sclera, and orthodontically straight teeth, the illusion that the original animal owner of the skull on the left looked like the near-human on the right was successfully conjured. Actually the evolution of Australopithecus sediba occurred in just a few short months. Millions of years were not needed. Homo sediba is such a nice looking guy; he actually looks like he could be my friend. The model is prominently displayed on the National Geographic website without comment or

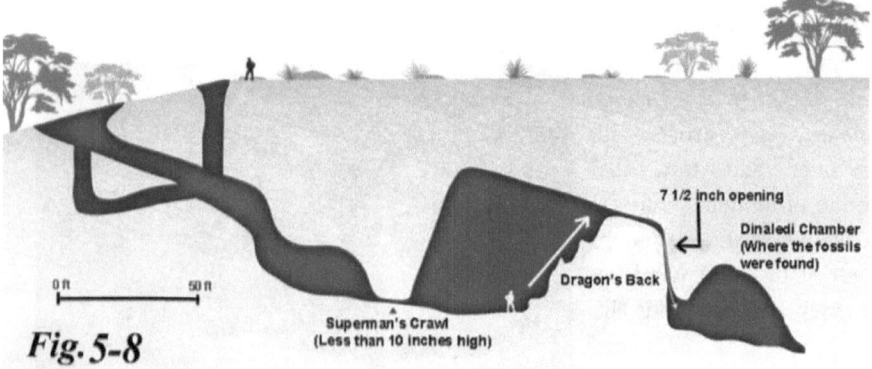

Fig. 5-8

disclaimer about the validity of the modeling. That would give away this illusion, and illusionists never do that. One National Geographic writer calls Australopithecus sediba "part ape, part human". The "part ape and part human" illusion was made solely by the sculptors and artists that constructed the model.[8-10]

The Strange Case of Homo naledi: In September of 2015, just as I was about to wrap up the last chapter of this book, newspapers throughout the world announced an amazing fossil discovery. A couple of spelunkers were exploring the inside the Rising Star Cave (Figure 5-8) in South Africa. They spotted some bones in a very remote chamber that had never been entered by modern humans. These bones became probably the biggest hominid fossil find ever. Lee R. Berger, a paleoanthropologist from Witwatersrand University, along with National Geographic, announced the discovery. The new hominid was named *Homo naledi*. I got the news from the Associated Press in my Orange

County Register newspaper. Of course I had to check out what was really behind the story to keep my book as current as possible.

The story of the find began in August 2013 in South Africa. It's recounted in the NOVA/PBS/National Geographic documentary titled *The Dawn of Humanity*. Most of the information in the following pages comes from this documentary, which you can easily access online. An ex-soldier, prospector, and adventurer, Pedro Boshoff, asked his friend Lee Berger for a job, as he was unemployed. Berger had known Boshoff for over twenty years. Berger really didn't have anything in the way of a job for Boshoff, but he told him to go out in the

Fig. 5-9

wild and look for hominid remains. Was Berger sending Boshoff out on a wild goose chase? In the documentary, Berger says most paleontologists that dig for fossils will never even find a single hominid tooth; and that 80 to 90 percent of the hominid fossil record is composed of just bits of isolated teeth. Hominid fossils are incredibly rare. Berger knows the chance of Boshoff finding even a tooth is infinitesimal. But Berger was so confident that Boshoff could find hominid fossils that he bought him a motorcycle so he could "move around out there". What a nice guy Berger is for buying Boshoff the motorcycle. And what a strange gift it was. Interestingly he bought Boshoff a street bike (Figure 5-9) that can be used only for street riding. Street bikes like Boshoff's are extremely heavy. They usually weigh over 700 pounds. They would be way too awkward to ride on dirt. If one tipped over, it would be very difficult to lift back up. If Berger wanted Boshoff to get out in the dirt and find fossils, why didn't he get him a dirt bike that could be used both on the street and on dirt? Was it nice for an employed dude like Berger to send an unemployed dude like Boshoff, who needs work and money, out on a search for fossils that Berger knows has a microscopic chance of being successful? Does Berger have a 6th sense? Does he know something no other paleontologist knows? It sure seems like he does because Berger told Boshoff to "go out and enlist your caving buddies, go underground, and see if you can find something... go to the most well-known places". A very odd suggestion, since the "most well-known places" would have the least chance of coughing up fossils. If fossils were present "in well-known places" they would have been discovered by now. But off Boshoff went, on his beautiful new road bike, straight to a "well-known" cave as Berger directed. There he ran into two spelunkers, Rick Hunter and Steve Tucker. He apparently didn't know either of them. Even so, he asked if they would search the caves for hominid fossils. Boshoff would cover the west area of "the Cradle of Humanity", an area that was thought to be where man first appeared, whilst Steve and Rick searched the east. Steve and Rick agreed to do the search. On September 13, 2013, Steve and Rick made their way into the Rising Star Cave (Figure 5-8), one they knew well. They went through a series of claustrophobic, dangerous, dark, and daunting openings, all because a complete stranger asked them to. I certainly wish I had Boshoff's power of persuasion. At what seemed like the very end of the cave was a tiny tubular opening called *Superman's Crawl*, which was less than ten inches high. They had to squeeze through, pushing with their toes; it was too narrow to pull themselves with their hands, or to crawl. Superman's Crawl was very claustrophobic; kind of like crawling through an underground drainpipe.

Superman's Crawl leads into a second large chamber with a 75-foot rising cliff at the end called *Dragon's Back*. Of course they climbed up the cliff. At the top was another miniscule opening. Steve crawled into the opening first. It went forward horizontally for about twenty feet. Then it made an almost 80^0 downward drop into another crevasse which was about 7½ inches wide, and as high as a four-story building. I'm getting acrophobic *and* claustrophobic just writing this. Of course Steve had no idea what was at the end of the downward 7½ inch wide opening. Except for his headlamp, it was pitch dark. As he descended, his feet were "dangling into empty space". He had no idea how far the bottom was. Picture yourself standing over a four-story building with another building 7½ inches away; in pitch dark. Would you squeeze yourself down in between the buildings to see what was on the ground? For all Steve knew, it could have been a 100-foot drop, and this could have been a fatal bit of spelunking, and his last. What would be the odds of finding hominid fossils at the end of this crevasse anyway? But Steve courageously kept moving. Did he know more than he lets on in this documentary? He squeezed down, and finally found a landing for his feet. What if the crevasse narrowed and he became stuck? Or what if it opened and he fell? Scary thought. Luckily it didn't narrow or cause a fatal fall. It opened into another cavernous chamber. He told Rick to join him, which he did. This is one of the biggest mistakes a spelunker can make. Both cavers were inside of a chamber that allegedly had never been entered by any man. Supposedly no person knew it existed. The entry was very miniscule, and difficult to navigate. Steve and Rick's cell phones wouldn't have worked inside that chamber. What if they couldn't climb back up the 40-foot crevasse to get themselves out? What if rocks came loose and trapped them? No person outside had any idea where they were. They could have died in that chamber, added their bones to the bone collection they were about to discover, and never been found. They did this amazing bit of caving simply because Boshoff, a complete stranger, asked them to. This is a real historic moment for evo-illusion. Did they know what they were doing more than NOVA intimated? At any rate, entering a hidden chamber inside of a cave that has a miniscule entry, without any person outside knowing where they were, was very dangerous. It completely goes against the *Safety Rules for Spelunkers*.

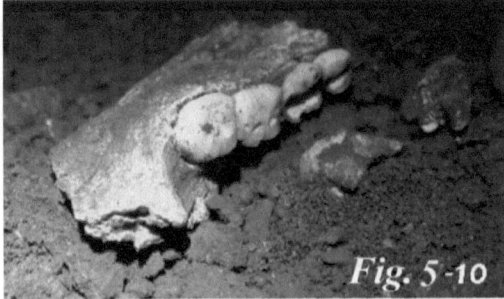

Fig. 5-10

Steve and Rick were shocked to find this last chamber was chock full of bones! In fact there were many long bones, and a broken piece of jaw with four teeth, just sitting there in the dirt. (Figure 5-10) Well, what would any normal spelunkers do? Of course pick up the bones and inspect them in hand. Then put as many as they could in their packs or pockets and hightail it out of that *chamber of death*. But they didn't. Instead they took photographs. For some reason Steve and Rick didn't collect any of the loose bones. After they took the pictures, they squirmed their way back out of the cave. They excitedly showed the pictures to Boshoff, who immediately declared them to be hominid. How did he know? There is absolutely no way of telling what the source of these bones was, but Boshoff knew they were hominid; for certain.

Well, off the trio went to Lee Berger's house. They showed the pictures to Berger, who also quickly declared them not only to be hominid, but "clearly hominid!" I am always amazed at how smart these paleoanthropologists are. I'm a dentist, and I couldn't tell what the source of the jaw fragment was just from the photo. But both Boshoff and Berger knew it was hominid, without a doubt. In fact Boshoff isn't a paleoanthropologist at all, which makes it even more amazing that he could tell. Did he have a preconceived notion, a *6th sense* like Berger? In saying it was hominid, Berger was also aging the jaw fragment. By declaring, "it's hominid", simply from looking at the photo, he's also saying it's millions of years old.

Now if I were Berger, the next thing I would have done is get right to that cave. I would give Steve and Rick some big bucks for their efforts. I would give them a bit of training so they could map and mark the locations of the bones. I would tell them to take good pictures, and then carefully pick them up and package them; and above all bring them out to me. After all, Berger didn't know Steve and Rick at all. Berger and Boshoff were too physically big to enter the fossil chamber. They couldn't even fit in the 10-inch tubular Superman's Crawl. Steve and Rick could have surreptitiously gone back, and removed every fossil for their own keeping or controlling. Fossils found in any country belong to that country, in this case, to South Africa. But Steve and Rick should have received the adulation and pats on the back, and been the stars of the NOVA documentary. It was they who risked their lives, and it was they who found the bones. Who knows what Boshoff might have done had they given the bones to him. In any case, he was far more responsible for finding those bones than was Berger. In point of fact, Steve and Rick are the true finders of the fossils, not Boshoff or Berger.

Even if Steve and Rick didn't go back and get the bones, if I were Berger, I would have been worried sick that word would get out, and an army of spelunkers would squirm their way into the fossil chamber, and remove them all. There were three people that needed to remain silent. Could Berger trust all three not to say a word to anyone? If they did, it could have been like the California Gold Rush. How about the South Africa Bone Rush? If I were Steve and Rick, I wouldn't even have told Boshoff about their find. I would have gone back and dug out every bone I could before Berger could confiscate my discovery; which is exactly what Berger did. Well, Steve and Rick didn't retrieve the bones on their own for some reason, known only to them. Instead this fascinating story turned into Showtime, devised and choreographed by Lee Burger. Berger became famous in the world of anthropology because of Steve and Rick's efforts. Steve and Rick will fade off into obscurity. Berger was the star of this historic NOVA documentary. Well, at least Steve and Rick were in it.

Berger took full command and control of the discovery, even though he had little to do with it. He contacted National Geographic who funded an expedition that Steve and Rick had *already* accomplished. Berger spent nearly two months assembling an international team of 60 scientists. I love the term *international team of scientists*. It sounds so impressive. Who could question an *international team of scientists*? He also needed some very thin people to fetch the fossils. Why did he need thin fossil fetchers since he already had Steve and Rick? Both were easily able to enter the fossil chamber. Was he worried Steve and Rick would claim the fossils as under their own control, as they could have and should have? Berger took over, and Steve and Rick didn't seem to mind. "I didn't know how I was going to find tiny people with extraordinary skills to work in the cave," Berger said. "So I did what my generation does. I turned to social media." Berger says he posted an ad on Facebook looking for people who are "skinny and not claustrophobic; they must work well with others. They must be able to drop everything in three weeks and come to South Africa—unpaid. Oh, and have a Ph.D. or masters in paleoanthropology." Berger had 57 applicants in a very short time. Most were young women. He chose six of the applicants to be cave-o-nauts. When the six petite volunteers got the word that they had been selected, each was excited beyond imagination. Each was certain this would be one of the biggest moments in their lives.

Berger, Steve and Rick, the *international team of scientists*, and the cave-o-nauts, gathered at the site. They set up a control center complete with tents and video hookups. They used a mile and a half of wiring so they could see

Fig. 5-11

inside the chamber, and observe what the cave-o-nauts were seeing and doing. Berger said the site was reminiscent of the opening of King Tut's tomb. I say it's also reminiscent of the control center for the first moon landing. It's Showtime!

The drama began. The petite women with their cameras squeezed through the tubular 10-inch Superman's Crawl, climbed the 75-foot cliff, and then squeezed into the horizontal fissure, and down the 40-foot deep 7½-inch wide crevasse (Figure 5-11) leading into the chamber where the fossils were. I do have to give these women, and Steve and Rick, lots of credit. The maze they had to go through to get to the fossils was beyond claustrophobic; very scary. I sure wouldn't have done it, so kudos to them all. The *international team of scientists*, Berger and Boshoff watched on the video monitor with bated breath. The excitement was palpable. And there they were! The images from the cave-o-nauts showed numerous hominid bones spread everywhere on the floor of the chamber. Strangely, Steve and Rick joined the cave-o-nauts in the chamber. Yes, there they were, on the TV monitor shown on NOVA. Which again begs the question, why didn't Berger simply ask Steve and Rick to be the cave-o-nauts? Why all the drama and the six women? Why did the women have

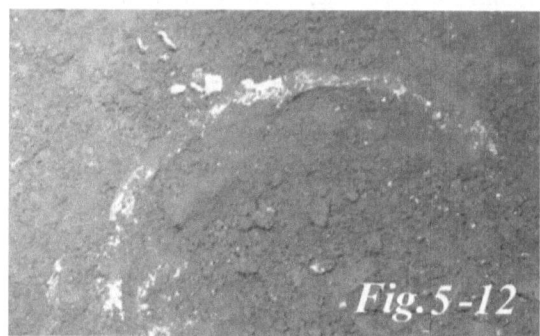

Fig. 5-12

to fly great distances and spend three weeks of their lives, away from their jobs and families? Why didn't any of them wonder why Steve and Rick didn't retrieve the bones? They were all so excited and giddy about the whole process; I doubt the thought ever entered their minds. Of course, *again*, these bones were declared to be those of hominids by Berger. This declaration was made almost immediately, because any old ape bones found anywhere on the African continent are immediately declared to be human ancestors. And this set of bones was no exception. They were also declared to be 2.2 million years old. What incredible conclusions. These bones are both from human ancestors, *and* 2.2 million years old. This is just a given in the venue of evo-illusions.

Virtually every news organization and every scientific society such as National Geographic and NOVA, and every museum of natural history in the world accepted both conclusions without question. In fact these astounding conclusions are never questioned; they are

Fig. -13

promoted. They are part of the illusion of human evolution. The cave-o-nauts spent three weeks retrieving bones. The skull (Figure 5-12) was buried with its half-round edge perfectly exposed in the surface dirt. How did it get buried in such a perfect yet strange fashion? It looks like someone buried it, then took his finger and wiped the top edge clean. It doesn't look like some random event happened that caused the burying of the skull. Maybe it was just some sort of crazy luck. Why wasn't the mandible (lower jaw) buried with the skull? How could one major part of the skull be buried, but not the other?

The last chore for the cave-o-nauts was to ever so carefully dig out the skull. (Figure 5-13) Why wasn't it retrieved first? Wouldn't it have been far more interesting to *first* see what story the skull had to tell about what kind of creatures left these bones? What if someone stepped on it and smashed it? If I were the paleoanthropologist, I would have taken the skull, by far the most important fossil, first. But that's just me. What was again amazing is that Berger was able to determine that the C-shaped object in Figure 5-12 truly was a skull. Can you imagine if that piece turned out not to be the skull of a hominid, after the *international team of scientists* and the six women had been working there for three weeks? It would have been one of the most embarrassing moments in hominid history. "Well folks, ah, we don't have a hominid as I thought we did, and uh, ha ha, ha…"

Beyond the obvious questions I have about the drama and choreography of the find, there are so many obvious questions regarding the fossils themselves. Firstly, how on Earth did the bones find their way into that chamber? Berger and the *international team of scientists* think the bones were placed there by Homo naledis as some sort of ritual for their dead members. The *international team of scientists* thinks the anatomy of the cave wouldn't have changed in the last several million years. The only way into the fossil chamber is through the cave opening. If they are correct, that means Homo naledi had to carry the bodies of its dead into the back of the opening chamber of the cave, (why didn't they just leave them there?) drag them through Superman's Crawl, into the second chamber, up the 75 foot cliff on Dragon's Back, through the first horizontal leg of the crevasse, 40 feet down the 7-½ inch wide vertically dropping crevasse, and into the fossil chamber. All of this had to be done in absolute and total darkness. A scientist on the *international team of scientists* asks the same question about how the bones came to be in the fossil chamber. He tries to answer it by saying,

*It looks like they got in there **because somebody put them there**. When you say that, it's a very controversial thing to say. There are no signs of predation. No predator accumulates only hominids in this way. There isn't a*

flow of material into the chamber. And this is where we leave it. Scientifically the best hypothesis we can say is they were put there. If this is true, the implications are far reaching.

The NOVA documentary goes on to try to build the illusion that Homo naledi actually did place the bones in that chamber as some sort of ritual for the dead. No other possibilities are even considered, not even the most obvious one. The *international team of scientists*, NOVA, and National Geographic, are incapable of thinking of any other possibilities except that this ape-like creature, with a brain the size of an orange, carried their dead through that unbelievable maze of crevasses, cliffs, and tunnels and into the fossil chamber; in complete and utter darkness. Berger speculates that this would have required light in the form of torches or fires placed at intervals in the cave. So the Homo naledi's who supposedly did one impossible activity, now, have to be credited with another impossible activity. A second illusion must be added to cover for the first, which is that Homo naledis discovered how to use and control fire. Just as one lie leads to another to cover for the first, one illusion leads to another to cover for the first. This ape, with a brain the size of an orange, is now credited with making torches or bonfires to light their way in the cave. And the evo-illusion expands.

Berger and his *international team of scientists* must protect this evo-illusion and this hominid at all costs. As preposterous as this entire story is, they will back it up with a straight face, just like any good illusionist will. To further convince its viewers that Homo naledi dragged their dead into the fossil chamber, PBS offered an article connected to their NOVA documentary on how animals take care of their dead which says,

"Apes have an understanding that death is irreversible", says primatologist Frans de Waal of Emory University. "When they see a dead group member, they're affected by it and they watch over it and sometimes try to revive it, touch it, and groom it... Mourning behaviors are common in chimpanzees. Animals will stop eating, observe a corpse in silence, and even carry dead infants for days or weeks. Moving bodies or burying them is not a typical primate behavior", de Waal says, "perhaps because most primates don't stay in one place for long. Now, if you live in a settlement where 50 other people live, you can't just leave a corpse there," he notes. "Elephants have occasionally been observed covering corpses—both elephant and human— though it's a less common response than tending to their dying and dead.

Indeed, elephants are one of the species that seems most affected by death, and will often stand quietly near carcasses, even from different families."

I'm trying to picture if elephants could carry their dead into that fossil chamber. I guess not. Does this dialogue convince you that Homo naledi dragged their dead into the fossil chamber? Were the bodies placed there by living Homo naledi's, as a ritual, when they died? Was Homo naledi more sensitive and intelligent than would seem possible? The fact that their brains are about the size of chimp brains should answer that question. If the bodies of dead Homo naledi's were dragged into the fossil chamber by living Homo naledi's, the next big question is who busted up the bones and scattered them all over the chamber? These fossil bones look like someone took a sledgehammer to the entire lot of them. If they were brought into the cave as dead Homo naledi's, their bones should be intact and positioned together. Skeletons should have remained fully assembled. There was no wind or rain or mud or stomping animals that could break up the skeletons and toss them all around the chamber. The fossil bones were completely isolated from the outside environment. It looks like whoever placed those bones in the fossil chamber also busted them up and tossed them all over. Interestingly, they had to be broken up years after the bodies were placed inside, when all of the tissues had been cleaned off the skeletons by bacteria and bugs. Two separate visits to each skeleton would have been required. And where were the other parts of the mandible? The jaw segment that was present was thick as if it was broken; not worn. The rest of the jaw couldn't have just dissolved, or disappeared, or eroded away, whilst the part that was found was unaffected. The missing jaw parts couldn't bury themselves. So where were they? Why wasn't the whole mandible *with condyles* present in that chamber?

Did Homo naledis squeeze through the cave's maze, in absolute darkness, and become trapped, and die inside the fossil chamber? Apes and monkeys don't do things like that, so this is completely out of the question. If that occurred, the skeletons would be fully intact. Do modern apes or monkeys place their dead in any cave, even ones that are open and easily accessed? Well, again, no. There really is no plausible explanation for how those bones came to be present inside that chamber. Actually there is only one other possible explanation. Were the bones salted there by a modern version of the Piltdown prankster? This question must be asked. This scenario has by far the greatest odds of being the explanation. It's never considered in the world of evolution's illusions. It certainly should be considered, after the many embarrassing hominids that were created out of faked evidence. A second question that must be asked is, who might have the greatest motive for doing such a thing? The list of possibilities is short, but, again, it's never considered. Evo-illusionists don't seem to be worried about being caught up in another fraud. True scientists should always be very careful not to let Piltdown Prank-man ever happen again. Evo-illusionists are not careful. They don't need to be, with so many people falling for every illusion.

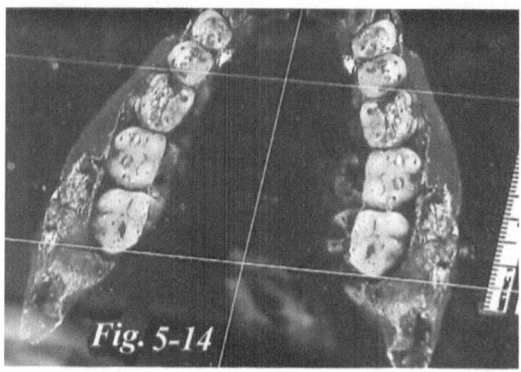

Fig. 5-14

The fossils of the hands and feet of Homo naledi are very complete, which is also very strange since the full skeletons are quite *incomplete* and missing many bones. We learned from other hominid fossils like Lucy that the hands and feet don't fossilize well because the bones that make them up are so small. Homo naledi broke that rule for sure. Homo naledi's fingers are curved palm-ward, kind of in a half grasping position, like those of most apes, and not like human hands at all. The thumb is longer than the first joint of the index finger, which is a definite characteristic of humans. Unfortunately, there are no fitted, naturally articulated, bony connections from the hands and feet directly to the skull, so it can only be *assumed* that they belong to the skull because they were found in the same chamber.

Homo naledi's jaw comprises two broken pieces of the right half of the mandible (lower jaw). One is the four-tooth fragment pictured in Figure 5-10. A team member from the *international team of scientists* put the two portions of the right side of the jaw together and photographed them. He then made a mirror

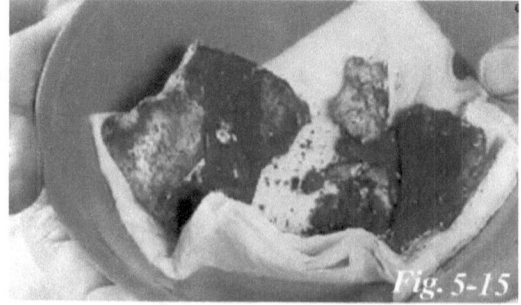

Fig. 5-15

image of the photo in his computer, which produced an image of nearly a whole lower jaw. (Figure 5-14) Only one problem here of several is that the condyles

Fig. 5-16

are missing, which are the ball-shaped heads that fit into the fossa or socket of the skull; which are also missing. Remember Piltdown Prank-man who was also missing condyles and sockets? So models and computer simulations had to be made to come up with the entire jaw, and skull. There is no possible way to determine if the jaw fits the skull. Even so, the fit is assumed by NOVA, NatGeo, and the *international team of scientists*, because the jaw and skullcap were found in the fossil chamber. Again, this certainly brings up memories of Piltdown Prank, whose condyles were removed from its lower jaw so it couldn't be determined if it fit the skull. Also, the computer artist had full artistic license to adjust Homo naledi's dental arch. There is no way of really

determining how narrow or wide the curve of the dental arch actually was. Was it narrow like an ape's, or spread like a human's?

Figure 5-15 is a photo of Homo Naledi's skull as portrayed on NOVA. This is all that was shown, so it must be surmised that that's all there was. If an entire skull were found, it sure would have been shown off; big time. So these skull fragments, and the two jaw fragments, and lots of plaster and imagination, produced Homo naledi's entire skull.

Figure 5-16 is a photo of Berger making goo-goo eyes with Homo naledi. He's obviously joyful

Fig. 5-17

about the find. More importantly, check out the frontal contour of Lee's own upper and lower jaws, from chin to nose. Notice how it's concave. Now check out the contour of Homo naledi's jaws. They protrude like any ape's jaws. Does Lee realize he's holding an ape skull?

The Homo naledi find was huge news. It was published in National Geographic, The New York Times, Associated Press, CNN, Scientific American, and most major newspapers and periodicals throughout the world. As I mentioned, NOVA made a two-hour documentary on the discovery. In fact, the November 2015 issue of National Geographic displays Homo naledi on its cover, (Figure 5-17) along with a big article on human evolution inside. Is Homo naledi really "almost human"? Is NatGeo pushing the evo-illusion of humans? Of course they are. The acceptance of Homo naledi in the scientific world of evolution was total. Essentially there were no skeptics or questioners. Well, except me. I question.

Think about this for a moment. Imagine if I spotted some bones like Homo naledi's in a cave. If I didn't retrieve the bones, but I took a few photos of them, and showed them to an anthropologist, would the anthropologist's first reaction be to round up a team of sixty international scientists? Would he put an ad on Facebook, asking for petite PhD scientists who would volunteer to fly to South Africa, and to crawl through tiny snaky cave openings? All of this when he doesn't really even know what kind of bones they are; or how many there might be? Would he transport the *international team of scientists* and the petite Ph.D. cave diggers to the cave? Would he set up tents and video monitors? Would he get National Geographic and NOVA involved? Would they be at all interested? Would everyone that watched what was found and viewed on the monitor by the petite cave-digging women be uber-excited? The odds of these being animal bones placed there by some human with an agenda, in recent history, or hominid bones, are about 100,000:1 in favor of recent human placement. It seems very odd that all of that money and effort went into searching out these bones before anyone knew what they really were and how many there were; before Berger even had a single bone in hand, does it not?

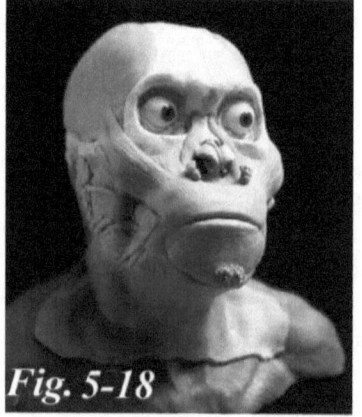

Fig. 5-18

National Geographic again hired evo-artist and sculptor John Gurche to make a model of Homo naledi. Gurche made some of the models of other hominids I discuss earlier in this chapter. He spent 700 hours on Homo naledi's model. He diligently placed and molded every facial muscle, placed the eyes, then the skin on a model of the skull. (Figure 5-18) He did everything any respectable evo-modeler could possibly be asked to do. And he did it well. In every case when Gurche had the choice of using ape characteristics or human characteristics, he chose human. Notice in Figure 5-18 how the braincase and forehead have already enlarged and evolved enormously, in only hours, not millions of years.

I put together a montage (Figure 5-19) showing Gurche's world famous model (left), the Homo naledi skull (center), and a model that I assembled using that skull for shape and form (right). I drew lines from the chin of Gurche's model, left in the collage, to the chin of the skull; and from the brow ridge on the model to the brow ridge on the skull. This allows for a size and proportion comparison of the skull and the models. Look at the enormous difference between the size of the

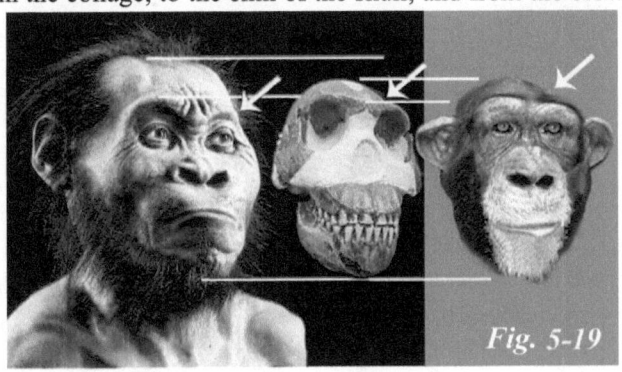

Fig. 5-19

cranium on Gurche's model and the size of the cranium on the skull. The model's cranium volume is about triple that of the skull. The skull has no forehead; Gurche's model has a large healthy one. Was Gurche told to add a cranium and forehead by NatGeo or whoever ordered the model? The arrow on the skull points to the location where the cranium descends and meets the middle of the browridge. On Gurche's model, the arrow points to where the cranium so obviously descends well laterally or outside of the browridge, like a human cranium. The brow ridges on Gurche's model are far smaller than are those on the skull. The model has human-type eyes with large sclera, a thoughtful human expression, human hair that looks like it needs barbering monthly, human skin, and human ears. Gurche explained how he gave Homo naledi a human-looking nose because the nasal bone protruded like it does in humans. But the entire nasal area of the skull was missing, and was filled in with white plaster by another modeler. Was Gurche unaware of this fact?

Gurche's model is completely inaccurate. It looks human when it should look ape, which is no doubt exactly what the Nat/Geo and NOVA evo-illusionists ordered and paid for. A photo of the model accompanied every article on this amazing find in every newspaper and journal. It was what I first saw in my Orange County Register.

Of course I had to make my own model. Unlike Gurche, I actually do use the Homo naledi skull as a guide. (right in Figure 5-19) I followed the

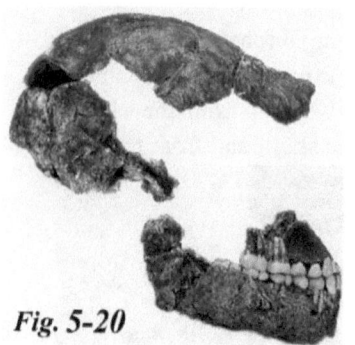

shape of the skull exactly, as Gurche should have. I gave my model ape eyes with no visible sclera, ape ears, and a furry ape face. This is a true version of what Homo naledi must have looked like. My model needed no barbering. The brow ridges follow those on the skull; so does the forehead, or lack thereof. And what do we have? An ordinary ape. Yes, Homo naledi isn't a hominid at all. It's an ape.

Fig. 5-20

Figure 5-20 is a photo of all of the skull pieces that were retrieved from the cave. Yes, that's all there is of the skull of an animal that is credited with being a human ancestor. That's all it takes. Find some broken up dirty primate bones… bingo, it's a human precursor. Notice the complete lack of nasal bones.

Just think. Tens of thousands of dollars were spent on Gurche's model. Whoever paid to have it done could have come to me for the job. I would've saved them a ton of money. Except mine wouldn't have supported the illusion of human evolution. My model would be an honest and accurate one that follows the skull shape and outline. It would have ape features instead of 100 percent human features, as Gurche selected. So, I guess NatGeo would have fired me on the spot. Gurche knows what he's doing, and how to keep his bosses happy. Can you imagine if Berger announced the finding of a new hominid, and it looked like my model? Like an ape, with all ape characteristics, like it should have? Berger would be a laughing stock. Because Gurche built his model to support the illusion of human evolution, Berger became famous in the world of paleoanthropology and NOVA viewers. Berger is a hero, and the star of his own NOVA documentary. I wish NOVA and

National Geographic came to me to make their documentary. Boy could I have helped them make an incredible and interesting documentary. But that'll be the day.[11-17]

One last hominid: Mr. Robustus: There is one more famous hominid, *Paranthropus robustus* (Figure 5-21) that I didn't discuss earlier; so to cover all the bases, here he is. Paranthropus robustus is prominently displayed at the Smithsonian National Museum of Natural History, so I just couldn't leave him out. The story of Mr. Robustus is the same, and just a predictable as it is for all of the other hominids. Paranthropus robustus was found in 1938 by Robert Broom. He was unearthed in Kromdraai, South Africa. He looked different than Australopithecus africanus, so Broom knew he had a new pre-human. Not a different ape species, but a pre-human; for sure. Broom collected many more

Fig. 5-21

bones and teeth. These guys were 3'6" to 3'9" tall. Their brain was the size of a squashed orange; except in the evo-artists versions. The Smithsonian says they inhabited the Earth from 1.9 to 1.2 million years ago. If you read the previous chapters, you're an expert at diagnosing hominid specimens, so I won't spend much time talking about Mr. Robustus. He had the same instant artistic evolution of a forehead, wider cranium, and human expression as the other hominids; in other words, the same gross misrepresentation. Check out the cranium on the skull, and compare it with the two evo-artist's rendition. Mr. robustus has a sagittal crest, which makes him an ape. Mr. robustus' cranium evolved to be ten times its size in just a few weeks, thanks to the evo-artists that painted these renditions. What more needs to be said? Both the photo on the right and the model are from the Smithsonian. Is the Smithsonian happy with such a gross misrepresentation? Apparently so.[18]

Ever since I started writing these chapters on human evolution, I've wondered. Aren't evo-illusionist organizations like NatGeo and the Smithsonian, and their artists, worried about having their illusions exposed as fakes? Hominid models are so transparent, the fakery is easily detected. But I guess they're not worried. What really amazes me is that they get away with their illusions, and that few people challenge or question. Don't they stay awake at night, worrying about getting caught?

To conclude this chapter, I assembled *The Hominid Dirty Dozen Facts*: twelve obvious facts that show the fraud of human evolution. Can anyone on Earth dispute any of them? Answer: no. But I welcome anyone who would like to give it a try. Well, here they are:

1. All supposed "hominid" skulls are partial ancient ape skulls. All have large brow ridges, small flat craniums, protruding jaws, "U" shaped dental arches, and miniscule to non-existent foreheads.

2. All evo-illusion artists make their models from ape bone finds using as many human characteristics and expressions as they possibly can to make them look like human precursors.

3. Zero hominid fossils demonstrate the gradual stepwise evolution of their craniums toward triple-sized human craniums and prefrontal cortices.

4. The diet of all primates is almost entirely vegetarian, made up of fruits, nuts, seeds, and leaves. For example, 97% percent of the diet of chimps is vegetarian. Human diets are very balanced between meat and vegetarian. If we came from an ape common ancestor, how and why does our diet vary so greatly from those of apes? A massive change in biochemistry is needed to make changes like this.

5. Finding a few anatomical or DNA similarities between an ancient ape fossil and humans does not prove that the ape evolved into modern humans. It proves only that similarities exist.

6. Documentaries made by respected scientific organizations such as National Geographic and PBS/NOVA will present human evolution from ape species as if it's a given. Reasonable challenges and questions are never addressed or mentioned.

7. Not one single modern ape species is capable of making tools or displaying human-like intelligence, which should give us an idea if ancient apes with similar brain configurations were capable of doing so.

8. Not one modern primate species out of 625 has the ability to communicate verbally or in writing to any degree that could be compared to what a pre-human should have been able to do.

9. According to evolution, the diversity of nature was created because of the splitting of species populations, usually due to massive geologic upheavals in the earth's surface and changes in its waterways. The newly split populations that inhabit new environments create new species. The populations that remained under the old environment usually didn't change much over tens to hundreds of millions of years. They exist today as modern species. But out of dozens of hominids, and dozens of chances, not a single hominid was able to survive only a few million

years to remain as a modern species. Six hundred and twenty five primate species survived as modern species. Humans survived to massively populate the earth with over seven billion people. But not a single hominid survived. An evolution scientist. told me the reason no hominids survived is that humans were more intelligent, made better tools and weapons, and had more advanced sociological structures. They out-competed the hominids. But the problem with this explanation is hominids and humans should have both out-survived and out competed primates. Hominids and humans both had superior intelligence, plus tools and weapons when primates had none. Plus they had far more organized social structures than primates. Hominids should have out competed primates. So this explanation doesn't fly. Hominids should inundate the earth today. If anything, of the three, primates should be extinct. Actually, if evolution is valid, the earth should be replete with humans, hominids, and primates. But 100% of hominids vanished. Which means they never existed in the first place, or in exponentially unlikely fashion, not a single hominid survived. Evolution's explanation for the diversity of nature completely collapses with primates, hominids, and humans.

10. The notion that Black Africans migrated out of Africa 60,000 years ago without the use of maps, compasses, boats, knowledge of where they were going, and an idea of why they were doing it, is ludicrous. To cross the Sahara Desert they had to cross at least 1500 miles of the most hostile and dry hot desert on Earth. The Sahara gets between 0 and 3 inches of rain per year. They had to also cross the Red Sea, which is 175 miles across in most places. Then came the Arabian Desert another impassable hostile desert that is 900 miles across. Then came the Persian Gulf, a sea of 225 miles in most places. Need I go on? The notion that Black Africans, or any group of humans 60,000 years ago, crossed these impassible barriers naked and barefoot is 100% not possible. Humanity didn't first form in Africa. And we didn't migrate out of Africa 60,000 years ago.

11. If you trace racial characteristics you will see that humans didn't originate in Africa and migrate out. Before man was able to travel via ship and horseback, sub-Saharan Africa was composed of 100% Black Africans. Europe was composed of 100% Caucasians. Asia was composed of 100% Asians. North America was composed of 100% American Indians and Hispanics. If Black Africans migrated out of Africa 60,000 years ago, there would be massive evidence of this migration. Black Africans should be mixed in with the Caucasians on Europe and with the Asians in Asia. But this is 100% not the case.

12. If you trace language characteristics you will find more evidence that humans didn't originate in Africa and migrate outward 60,000 years ago. There is zero African language influence in any European language, or in any Asian language. No African words are imbedded in any language outside of Africa.

Human evolution from apes is nothing but a scientific illusion. At this point in time, there is no possible objective and scientific explanation for the appearance of man on Earth.

Chapter 6

Demographics Tell the Tale

Polar bears did very well in the warmer times. They didn't die out at all; they didn't die out in the last 10,000 years, nor during the previous interglacial, nor the one before that. So, they're just used as a deceitful heartthrob; you know, to pluck your heartstrings because the polar bears might die out. -Piers Corbyn

The average rate of the doubling of any population over time can be a very handy tool used to determine if science's age estimates of any species are correct. The reason average doubling times are of such intense interest is because, knowing the birthrate, average life span, and average generation length of any species yields the average population doubling time of that species. Then, using current population figures, the approximate length of time a population of any species that has existed on Earth can be reasonably calculated. Average population doubling times provide a way to check other methods used in estimating the length of time a species has existed.

The formula for calculating the doubling time of any entity that increases in number over a period of time, be it money, or populations, or anything, is called *The Rule of 70*. It looks like this:

dt=70/r

Where dt = doubling time, and r = the rate of growth as a percentage. For example, given Canada's net population growth of 0.9% in the year 2006, dividing 70 by 0.9 gives an approximate population doubling time of 78 years.[1]

My human population research got me thinking about population doubling of animals that are on the top of their food chain. What would population-doubling studies look like for them? Comparing and contrasting population studies of an animal that is at the top of the food chain with humans who are at the top of their chain as well should be very educational and telling. For my animal study I selected the most perfect predator I can think of, the polar bear. Polar bears are at the absolute top of the food chain in their habitat. No animal can threaten them. A rare exception is the walrus that can kill polar bears by throwing their heads back and stabbing forward with their huge tusks. An event like this is a rarity. On land or ice, polar bears win easily against walruses. Usually they win in water as well. Man has been an occasional and not significant predator of polar bears. Polar bears rarely kill other polar bears. There are few significant polar bear diseases that would be population reducers. Polar bear mothers are dedicated protectors of their cubs. If you get near her cubs, expect an attack in short order. Polar bears live an average 15 to 18 years, although biologists have tagged a few polar bears in their early 40s. In 2008, scientists estimated world polar bear population at around 25,000. Polar bears are the perfect top-of-the-food-chain species to do a population doubling study on. Does population doubling verify what science has to say about how long they've inhabited the Earth?

Polar bears have one of the slowest reproductive rates of any mammal. Females breed yearly during their fertile years. Female polar bears have their first set of cubs at around 5 or 6 years old. The ravages of old age prevent females from having litters four to five years before they die. Litter sizes can

range from one to four cubs, with an average of two cubs. They're born in a snow den, which the mother digs, between November and January. Polar bears don't hibernate in the strictest sense of the word. All true

Fig. 6-1

hibernators hibernate annually like clockwork. They experience a marked drop in heart rate and body temperature. They hibernate whether they have young or not. Adult polar bear males and non-pregnant females don't den up at all. Mamma bears remain in the den, nursing their cubs until April.

The mortality rate for cubs is estimated to be 10-30%. Since the average lifespan of a female bear is 18 years, and they cannot reproduce while they're under about five and over about thirteen years old, females should produce eight or nine litters in their lifetime, yielding sixteen to nineteen cubs. Instead of calculating that female polar bears have eight or nine litters, I will very conservatively use five litters per female for my population calculations. This will generously take in any other unknown and unseen population reducers that might occur. [2-5]

Since polar bears have an average lifespan of 18 years, two must be subtracted from polar bear population growth every eighteen years. Using the high estimate that 30% of cubs do not survive, their population should rise 2½ times every 18 years. (Five litters per mama times two cubs per litter = ten cubs; minus the two parents who will die of old age, and the 30% or three cubs that do not survive.) Very conservatively, two parent polar bears will turn into five every eighteen years. So it can be concluded that, according figures that are far more conservative than those given by polar bear experts, polar bear populations should double about every 15 years or so. Polar bear populations should rise from the current 25,000 polar bears to:

50,000 in 15 years;
100,000 in 30 years;
200,000 in 45 years;
400,000 in 60 years;
800,000 in 75 years;
1,600,000 in 90 years;
3,200,000 in 105 years;
6,400,000 in 120 years;
12,800,000 in 135 years...

Fig. 6 -2   Rendering of an Ursavus by Margaret Lambert

As you can see, in the year 2141, there should be over 12,800,000 polar bears on Earth. One hundred and fifty years later, there will be over 13,000,000,000 (13 billion). In the not too distant future the world will be inundated with polar bears. Of course, this isn't a possible outcome. How can this absurd figure be explained? Taking a closer look at polar bear populations and the length of time modern scientists have assigned to their time on Earth makes things even more interesting.

According to evo-illusionists, *ursavus* (Figure 6-2) inhabited Asia 22 million years ago, and was the common ancestor of all bears. It looked pretty much like an ugly pitbull dog. How anyone could determine that ursavus evolved into a bear is anyone's guess. But evo-illusionists have to pick some species, so why not ursavus? Just as with birds and whales, there just aren't any early species that looked much like a modern bear, so their pickens are slim. Which scientist is going to argue the point, considering their jobs and grants would be threatened? So ursavus is the common ancestor of all bears by declaration, and don't ask questions. In searching out fossil evidence of bear evolution, I was surprised that there was little information. Bears supposedly evolved fairly recently in geologic time. There are no bear fossils to study. I was unable to find much in the way of bear precursors except for Ursavus; a pretty pathetic example.[6]

According to evo-illusionists, about 1.6 million years ago, grizzly bears, the descendants of ursavus, migrated to North America by the ever-mysterious Bering Land Bridge. A group of Grizzly bears became isolated and adapted to life on ice becoming polar bears, between 400,000 and 500,000 years ago.

According to Polar Bears International:

*Estimates of when **polar bears began to split from brown bears** continue to change as geneticists look further into the **polar bear genome**. The most recent paper now puts the evolutionary time frame at around 400,000-500,000 years. After beginning to branch off from brown bears, the polar bear's ancestors underwent a series of rapid **evolutionary changes** (in less than 20,500 generations) in order to survive in the harsh conditions of the Arctic. The bears **adapted** to a life of hunting seals and surviving extreme cold.*

Other evo-illusionary estimates say polar bears appeared 300,000 years ago. So we can safely say the range of the age of polar bears is 300,000 to 500,000 years. Using the most conservative estimate of the existence of polar bears, 300,000 years, and if they doubled their population every 15 years, which again is extremely conservative, would mean polar bears doubled in population 20,000 times since their beginning 300,000 years ago. (300,000 years/15 years=20,000) If only two polar bears existed 300,000 years ago, one male and one female, again, two being the minimum needed for procreation, and if those two doubled in numbers every 15 years, there should be over 4 x $10^{6,020}$ (2 times itself 20,000 times, or $2^{20,000}$) polar bears on Earth today. If you are not familiar with this kind of numerology, $10^{6,020}$ is a one with 6,020 zeroes after it. One trillion is a one with twelve zeroes. This means that today there should be enough polar bears to fill this universe, and trillions of trillions of other universes, solid with polar bears. Of course this cannot be the case.[7,8]

Consider if polar bear populations doubled every 300 years instead of every 15. A 300-year doubling time just doesn't fit any of the numbers given by polar bear experts. But for the sake of discussion, let's see what a 300-year doubling time looks like. This means that if there were the required minimum two polar bears on Earth 300,000 years go, it would take 300 years to have a population of four. In six hundred years there would be eight... and on and on. Those eight would grow in numbers to 10^{301} polar bears today. There are approximately 10^{80} atoms in the universe. A 300-year doubling time would produce exponentially more polar bears than there are atoms in the universe.

If polar bears truly existed on Earth for at least 300,000 years, and their current population truly is 25,000, the average doubling time would be 300,000 divided by less than 15. Fifteen is the number of times they would double to reach the current population. If 2 were doubled fifteen times the resulting population would be 32,768. On average polar bear populations would need to double only about every 20,000 years. Which means that if only two polar bears existed 300,000 years ago, there would be, again on average, only four polar bears 20,000 years later; and only eight 40,000 years after that; and on and on until the current population is reached. Is that possible? Of course not. No species would survive such an absurd population growth rate. Polar bears would have quickly become extinct.

Remember, polar bears have lived all this time at the top of the food chain, with no predators. Even humans weren't polar bear hunters for 99.999% of polar bear existence. To try to reach reasonable figures, and really tame these numbers down, if there were the minimum 2 polar bears in existence *only* 10,000 years ago, with a 15-year average population doubling time, there should be 3×10^{200} polar bears in existence today. (2^{666}) Again, there are approximately 10^{80} atoms in the entire universe. So you can see how impossible even a 10,000-year origination time for polar bears is. If polar bears are 10,000 years old, their modern population should be 10^{120} times greater than all atoms in the universe; a complete absurdity.

What number actually does work? Starting with the minimum two polar bears 200 to 225 years ago, if they doubled in population every 15 years, the resultant population would be the 25,000 polar bears that exist today. Of course the notion that polar bears have only existed on Earth for 200 to 225 years is nothing but preposterous. In fact all population studies with polar bears are absurd. But they must be explained for evo-illusion's story to be valid.

Anyone who wants to challenge my numbers would most likely say that there are lots of factors that would kill polar bears, which would keep their population way down. These factors would prevent the normal population doubling that I'm talking about. But remember, I've very generously accounted for all population reducers in my math. I calculated a 30% death rate for cubs, the highest estimate from several polar bear expert sources. I reduced the accepted number of cubs per female from 18 to 10. I reduced the number of litters from ten to five. My figures do generously account for *all* early deaths by *any* means. I way overestimated polar bear death rates.

If disease and starvation killed millions of polar bears, there certainly would be polar bear skeletons by the billions all over the Arctic tundra and seas. There would be polar bears that looked like they just walked out of Nazi prison camps; emaciated, weak, sickly. But that just isn't the case. Think how many polar bear carcasses would be lying around if starvation were a big problem. If polar bears died by disease and starvation over hundreds of thousands of years, and that were the reason for such a small modern population, there would be immense amounts of carcass and skeletal evidence. But that evidence just doesn't exist. Meat eaters and bacterial deterioration would eliminate the meat on polar bears that died for any reason. No meat eater eats bones; well, with the exception of dogs. Arctic seas should be filled solid with polar bear bones. They are not. Mankind should right now be having an immense problem clearing lands of polar bear skeletons, and the skeletons of other "top of the food chain" predators as well. Tractors should be busy full time clearing lands that are covered by polar bear bones. John Deer *bone movers* should be the largest selling tractor type on Earth. Where are the skeletons that would show the reduction in populations that would give a hint of the number of years polar bears have populated the Earth?

Is there an answer to this puzzle? What is known for certain is the size of polar bear litters, their life span, the survival rate of cubs, and the number of litters per mamma bear. Females have litters annually, and male sexual prowess will not turn off and on to accommodate such a low population of polar bears. They just don't cease procreating. If they did, they would be completely extinct in a very short time. Polar bears breed annually, like clockwork. No factor is going to stop that process. Have polar bears existed on Earth for only a couple of hundred years? Of course that's an absurd notion. So what is my answer to this paradox? I wish I had one. I, for the life of me, cannot come up with any kind of solution to the *Polar Bear Puzzle*. Could polar bears truly be 300,000 years old? I don't think so. These numbers don't fit any kind of demographic scenario. Actually neither does a 10,000-year existence. No, I do not believe in a young Earth. The Earth gives way too many clues that it's billions of years old. So I guess I'll have to leave this fascinating puzzle unanswered and think about it later.

I do have an idea for a great movie, though. Just imagine if polar bears procreated at the rate that scientists say they do. There would be trillions of polar bears covering North America in a very short time. Humanity would be inundated with these unstoppable white monsters! We wouldn't be able to make enough bullets to kill them off. They would break into people's houses at their whim, and wipe out entire families. They would eat people and every other animal they could get their gigantic paws and teeth on. They would decimate the North American continent and start moving into Central America. South Americans would be widening the Panama Canal to prevent polar bears from inundating their continent. The army and marines would have to be called out to fight the oncoming white monsters from the north. Brad Pit could lead the military. I think he does a great job in war movies. Scientists would confer on how to stop the polar bear onslaught. Of course other top actors would be my scientists. Hey, maybe Bruce Willis and Dustin Hoffman. They may need to plan how nuclear weapons could kill millions of polar bears; but millions would only be a drop in the bucket. I would need a beautiful but very intelligent actress who loves polar bears, and decries the killing of these magnificent creatures; some sort of Green Peace kind of person. Hey, maybe Angelina Jolie. Perfect! She, of course, would wander out and try to communicate with the attacking polar bears, and nearly get herself killed doing so; at least several times, to make the movie extra exciting. Well, I'll keep thinking about the solution to this puzzle, and possibly start writing a script for *Monsters From the North*. No, that's pretty corny. Maybe a better, a more intriguing title, would just be *Polar*. That should keep me busy. I really think I have a potential hit on my hands. And maybe an Academy Award! Gosh, how exciting.

Polar bear demographics are a fascinating study. But what about humans? I wrote about human demographics in *Evo-illusion*, but here is a brief synopsis. Can human demographics give us an idea about how long humans have inhabited the Earth? Evo-illusionists say modern humans have populated the Earth for the last 200,000 years. So let's put their number to the test. To give you an idea about using doubling times to test the age of humanity on Earth, consider the growth of the population of the Earth in the twentieth century. The population of the Earth was 1,654,000,000 in 1900. The population in 2000 was 6,100,000,000. Earth's population doubled almost twice in the twentieth century alone. During the twentieth century we had, for the first time in history, very successful forms of birth control. That alone is a powerful population reducer that didn't exist before the twentieth century. We also had aids and flu epidemics that killed millions, World War I and World War II, major holocausts, famines, natural disasters... These and other population reducers *should have made* the twentieth century population-doubling times exceedingly long. But, in the twentieth century, the world population doubled in a very short time: about every 53 years.

Fig.6-3

In the fourteenth century, the bubonic plague (Figure 6-3) caused a period called the *Black Death*. It was one of the most devastating pandemics in human history, resulting in the demise of an estimated 75 to 200 million people. It peaked in Europe around 1346–53. Human population doubling time for that exceedingly tragic and unusual period should have been historically long; and it was. According to demographers, the population of the Earth took 450 years to double from 1300 to 1750. Considering the long doubling time during the bubonic plague, and the short doubling time of the twentieth century, the average human population doubling time should be somewhere between 53 years and 450 years. If mankind is given a very conservative average population doubling time of 200 years, evo-illusion's 200,000-year-old man can be tested. Starting with a population of two, the minimum needed for procreation, for mankind to have an age of 200,000 years, with an average population doubling time of 200 years, means the Earth's population would double 1,000 times. (200,000/200 = 1,000) Two doubled one thousand times ($2^{1,000}$) should yield a modern population of 1.1×10^{301}, a number so high it doesn't exist in reality. There is no entity in the universe that exists in that quantity. That number of people would fill an immense number of universes solid with the people. So either a 200,000-year-old man, or a very conservative 200-year doubling time just isn't possible.[9,10]

What if 450 years were used, the longest doubling time in recorded history; the one that occurred during the Black Plague? A 450-year average population doubling time would yield 444 doublings in 200,000 years. (200,000/450=444). Two doubled 444 times (2^{444}) would result in a modern population of 4.5×10^{133}, a number so high it again would fill uncountable universes solid with people.

Actually, if the Earth's population doubled 32.5 times ($2^{32.5}$), the resulting population would be 6,074,000,000, which is the current population of the Earth. The average doubling time would then be $200,000/32.5 = 6,153$ years.

Which means the family of the original male and female would have to wait an average of 6,153 years to number four people; 12,306 years to number eight; 18,459 years to number 16; and on and on. Something is very demographically wrong with a 200,000-year-old human.

Scientists say mankind migrated out of Africa 60,000 years ago. This can also be demographically tested in the same way. Using the scientifically accepted human population 60,000 years ago of 10,000 people, and using the very long and rare 450-year population doubling time that occurred during the Black Death as an average doubling time, means that the population must double 133 times in the last 60,000 years. (60,000/450=133). Doubling 10,000 people 133 times (2^{133} X 10,000) results in a modern population that is *105 trillion* times greater than the actual current world population. Yes, there should be at least *108 tredecillion* people on Earth today, a number most people have never heard of. One hundred and eight tredecillion is a 108 with 42 zeroes after it. It looks like this:

108,000,000,000,000,000,000,000,000,000,000,000,000,000,000

In reality, the average doubling time that would produce the population of the Earth today from 60,000 years ago had to be a bit over 3,000 years (60,000/3,000=20 doublings; 2^{20} X 10,000=10 billion). Three thousand years is about as long ago as when King Tut ruled Egypt. To put these numbers in perspective, if there were a tribe of 100 humans, and they lived at the time of King Tut, they would number 200 today. Of course this is not a possible scenario, which means that humans could not have migrated out of Africa 60,000 years ago. How do I account for these strange demographic numbers? I don't. I cannot. But they do indicate that something is very wrong with the numbers given by evo-illusionists.

In the next chapter I will discuss the "out of Africa" migration that one can see so commonly displayed by maps in anthropology textbooks and documentaries. Try Googling "human migration maps". Select "images", and you will see hundreds of them. It's a completely accepted scientific perception that humans evolved in Africa, and migrated outward to all continents starting 60,000 years ago. Do demographics support evo-illusion's storyline? Actually, population studies turn that notion upside down. The next chapter will take a good scientific look at human migration.[11-14]

Chapter 7

Are Those Maps with Arrows Real?

Science is a way of trying not to fool yourself. The first principle is that you must not fool yourself, and you are the easiest person to fool.
– Richard Feynman,

According to the given scientific timeline for the existence of humanity, humans first appeared in Africa 200,000 years ago. Well before that, around 900,000 years ago, a group of hominines had completely had it with Africa, and wanted out. Just imagine what population figures would be with a 900,000-year-old human! In any case, they must have been just disgusted with Africa for some unknown reason. Africa was so thinly populated, and so full of natural resources. It had possibly the best climate on the planet. The migrating hominines could have oh so easily migrated to thousands of locations in Africa and had everything they could have possibly wanted in the way of food resources and living conditions. But supposedly they wanted out. According to paleo-illusionists, these aggravated hominines started their migration somewhere around central Africa. They migrated north, then east. Their goal was to find their way to a better place, and that place was going to be what is now the Peking area of China. Their first major barrier and challenge would have been crossing the Sahara Desert. Can you imagine running into a group of

ape-like pre-humans crossing the Sahara? (Figure 7-1) Wouldn't you wonder what the hell they were doing? What a sight that must have been. The disarray must have been unbelievable. Gad.

Of course, they had no shoes or clothing; or sunscreen. I wonder why they didn't burn their feet to ashes. I also wonder what they did for water in the Sahara. They certainly had no containers to carry water. Even if they did, they

Fig. 7-1

couldn't carry enough to cross the Sahara's 2,000-mile expanse. The Sahara became desert 2 to 3 million years ago. To answer the challenge of how hominines crossed the Sahara, geologists say during different periods it supposedly morphed from desert to savannah and back to desert. How do they know this? Well, from cave drawings that display the Sahara as a plush savannah of course. That's giving an awful lot of credit to ancient cave dwellers, but when the evidence is so scarce, what else can evo-illusionists do? OK, so the daring group of hominines that successfully made the trek must have crossed the Sahara when it was savannah. Actually, savannah or desert makes no difference. It was still an unbelievable and deadly expanse to cross.

The Sahara desert comprises eight percent of the world's land area. It's so immense that one could place the entire continental United States within the Sahara and still have a few thousand square miles of desert left over. Oases formed by underground aquifers cover over two percent of the Sahara. It's made up of about 30 percent sand and 70 percent rocky plateaus and gravel plains. The Sahara has produced some of the hottest temperatures on the planet. In fact, the all-time hottest temperature ever recorded was 136° F, in Azizia, Libya, in 1922.[1] One scientific site said:

The desert of Sahara is supposed to be at least 2.5 million years old. Studies made on the fluctuations of humidity in the Sahara during the last 40,000 years revealed that the borders of the desert moved sometimes southward and other times northward and in particular periods, the desert disappeared completely, the sand dunes being replaced by wooded savannas, like those found today in eastern Africa.[1]

There's an obvious problem with this scenario. For the Sahara to cyclically morph into a savannah and then back to desert, it repeatedly needed an unimaginably immense amount of soil. Where would this incredibly immense amount of soil needed to form savannahs come from? It has been reported that the average depth of the sand in the Sahara is approximately 500 feet. The sandy part of the Sahara comprises about one million square miles. The tops of the dunes can reach a height of 1,000 feet from bedrock. So the sandy area of the Sahara is composed of over 100,000 cubic miles of sand. For the fun of it, this translates into 1,504,000,000,000,000,000,000,000 or one septillion, five hundred and four sextillion grains of sand. Where did this sand go when the Sahara morphed into a savannah? How exactly were the sand dunes "replaced" by the rich soils required to form savannahs? How did 100,000 cubic miles of sand return for the next desert cycle? How did the soil and sand replace each other? This is an unimaginable scenario. Scientists know this happened because of cave artists that painted savannahs on cave walls. Just think, those ancient cave paintings trump all logic. Well, apparently the cycling of the Sahara desert happened numerous times fortunately for hominines 900,000 years ago who wanted to cross it and make their way eastward.[2-4]

For those thrill-seekers who desired to leave the beautiful greenery and plush environment of the south half of Africa for a nightmarish trek north, the Sahara stood as an impassible barrier; even if it was a savannah. Migrating north was actually a much more daunting task for the first hominines than was going to the moon for modern man; and actually far more dangerous. The rest of this chapter takes a critical look at the timelines and those *maps with arrows* proffered by evo-illusionists.

The earliest migration out of Africa occurred 1.75 million years ago. Some sources say the first hominid to leave Africa and trek into Eurasia was Homo ergaster.[5] As you can imagine, the estimated dates of all migrations vary greatly. Different respected sources give very different timelines. So in discussing what modern evo-illusion has to say about human migration, only very approximate dates can be used. No matter what, all hominines that left Africa had to cross the Sahara Desert, a daunting and deadly task. Hominines that ventured to Peking had to have started their trek between 1,750,000 and 900,000 years ago. The entire notion that the sand of the Sahara was replaced

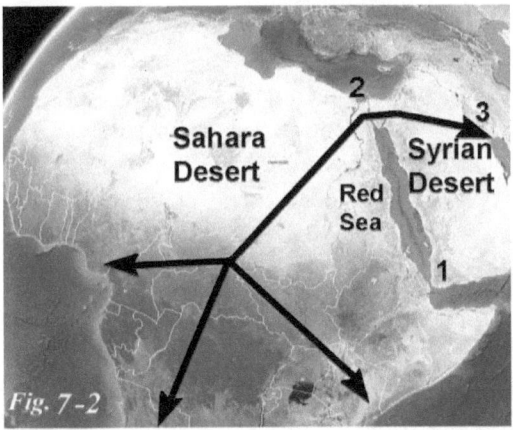

Fig. 7-2

with soil and grasslands on a cyclical basis seems preposterous on its face; but, for the sake of discussion, let's imagine that did happen, and a group of hominines was able to trek across the Sahara. Their next barrier would have been the Red Sea, which has an average width of about 160 miles. The hominines that wanted to cross it had two possible crossing sites. On the south of the Red Sea is a narrow opening about 20 miles across called the Mandeb Strait (1 in Figure 7-2). The notion that hominines that lived over 900,000 years ago could build boats and make a 20-mile water crossing is absurd, so that must be eliminated as a possibility. The only way the Peking-bound hominines could have crossed the Red Sea is at the north end where there's a land bridge that is now part of the Suez Canal. (2 in 7-2) Without a compass, maps, or Google Earth, the trek from the very green and plush southern half of Africa to the area where the Suez Canal exists would have been nothing short of a miraculous undertaking. Any hominines, with zero knowledge about the geography of Africa, trying to cross the Sahara would have wound up going in circles; they would have all died. But, let's assume these are incredible never-say-die-and-very-smart hominines, and they did make their trek to the land bridge in the north. The next barrier is again unthinkable. They would have had to cross the Arabian Peninsula, which makes up most of Saudi Arabia. The Arabian Peninsula is 1,000 miles wide, and almost all desert. I really don't think we can count on the same repeat addition and removal of sand and soils that made cyclic savannahs of the Sahara. Crossing a desert-like the Arabian Peninsula without a plentiful supply of water, a map, a compass, or GPS would have been impossible. The hominines would have again been instantly lost, just like they would have been in the Sahara. There would have been no food or water for the trekkers. They would have trekked in circles and soon died of thirst or starvation. No intelligent modern human could survive that journey on foot without modern technological help. How would ape-like hominines navigate the Sahara and the Arabian Peninsulas 900,000 years ago? Why would they even think of trying? They couldn't have possibly had a destination goal in mind, since they had the

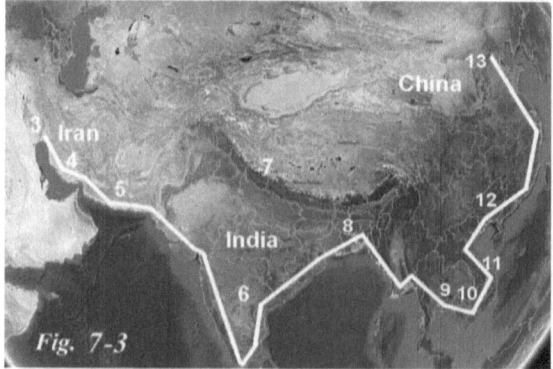

intelligence level just above that of an ape. Oh yes, I forgot. They wanted to visit, and possibly live in Peking. Any hominines who attempted to cross the Sahara Desert, the Red Sea, and the Arabian Peninsula 900,000 years ago would have died in short order. But, again,

Fig. 7-3

let's use our super assumption powers, and assume they were able to do so. These intrepid hominines would have wound up at the top of the Persian Gulf, where Kuwait is now located. (3 in Figure 7-3) Rather than cross the Arabian

Peninsula, the intrepid hominines certainly could have stayed along the shoreline of the Red Sea, and trekked around what is now Saudi Arabia, Yemen, and the United Arab Emirates. They would have been close to seawater for the entire circumnavigation. It might have been a bit cooler, and they could have taken a swim if they got too hot. Since primates cannot drink seawater, what did they drink? Circumnavigating the peninsula would have required trekking over 4,000 miles, a daunting task for barefooted hominines without food or canteens. In reality, the only travel choice that would have allowed them to survive was to cut straight across the Arabian Peninsula. Actually the chance of every member of the group dying with either route is 100 percent. But let's hope and suppose they did make it to the area of Kuwait, still intact and alive. Next they had to navigate the coastline of Iran (4 on Figure 7-3), and Pakistan (5), a distance of about 1500 miles. Next came India (6), which is crossed by the Himalayas (7), a mountain range and a barrier that would be more daunting than the Sahara. To avoid the Himalayas, they must have followed the coastline of India, a trek of about 3,000 miles. I bet by this time they were getting the "When are we going to get there Daddy?!" questions from the kids that all parents detest on long trips. I bet their feet were killing them! The big question here is why did they keep going? Weren't there some rather interesting and comfortable places in Iran or India that might have piqued their interest for a place to settle? I guess not, since there are no signs of remnants of these very brave sub-humans along *any* of their route. These guys wanted Peking, and nothing was going to stop them. So they kept on trekking.

Next they had to follow the coastline of Bangladesh (8), which is almost 300 miles long; then the coastline of Myanmar (Burma) (9), which adds another 900 miles to the trip. Burma can be a pretty miserable place, from what I've seen watching World War II films. Hot, sticky, mosquitoes, monsoon rains… terrible for trekking. They then had to follow the coastlines of what is now Thailand, (10) Cambodia, (11) Viet Nam (12), and southern China. They didn't just quit when they arrived at the southern coastline of China. They continued northward, until they reached their goal: Peking! (13) Well, the area around what is now Peking. And there they settled; and they became Peking man. The final leg of their journey from Burma to Peking was about 4,300 miles. Of course all of the mileage figures are dependent on the hominines not getting lost and going in circles on any parts of their trek, which would have been easy to do. Remember, no person had ever mapped their route, and they had no compass or other means of navigation. Of course, they had no idea where they were going since no human nor pre-human had ever been there before. Boy, it sure must have been a good feeling when they finally made their goal. I bet they were pooped. What is really strange is that the intrepid hominines, AKA Peking man, bypassed Hong Kong, which has very mild temperatures and a great climate. Wintertime averages in Hong Kong can range around 61^{0}F, whereas in Peking (now Beijing) the wintertime temperatures can reach 15^{0}F. What on earth were these guys thinking? But it was what it was; for some reason they chose Peking to settle down. I really wish I could have been there to give them some advice, and maybe steer them back to the area around what is now Hong Kong, but that's just a pipe dream. They probably wouldn't have listened to me anyway. Everyone makes mistakes. The British Museum of Natural History had a drawing made (Figure 7-4) depicting and celebrating one of the greatest feats in the history of sub-mankind, or even mankind; probably the greatest. In celebration of their incredible journey, it appears they ran around naked, feasted on deer, and busted up some rocks. What a bunch of party animals! I say hats off to Peking man; and kudos as well for making an impossible journey possible. I certainly hope they figured out a way to make clothing. They would have frozen their butts off dressed like that in their very first winter in Peking. Who knows how many years it took, or how many generations were dedicated to their trek from Africa. If the trek did actually occur, I'm sure it was many years. Would hominines that break rocks to make tools and live naked in caves as shown in the above drawing be capable of a trekking thousands of miles from Africa to Peking? Would they want to? Evo-illusionists say yes; I say no. I say the trek from Africa to Peking over 900,000 years ago is nothing but an evo-illusion. It would have killed any hominid who tried.

To prove this group of hominines trekked all the way from Africa to Peking, paleoanthropologists dug up four teeth and a skullcap. (see Peking man in Chapter 3) All of this evidence from 900,000 years ago is so inspiring. I just can't imagine the

Fig. 7-4

hardships these daring hominines encountered. Peking man may have evolved into every Asian person in the world. Or maybe not. One or the other. Either they did or they didn't evolve into Asians, but evo-illusionists have it narrowed down to those two choices.[5]

700,000 years ago: The second major migration out of Africa occurred. Onward to Java. Over two hundred thousand years after the arrival of Peking man in eastern China, word must have gotten back to a group of Homo erectus in central and southern Africa. News travelled slowly in those days. Maybe African Homo erectus found out what a terrific place Peking was. So another group set out to seek their fortunes and take the same route as did the first group of intrepid travelers. Once they reached southern Burma (Figure 7-5), they trekked along what is now the west coast of Thailand for 300 miles (1).

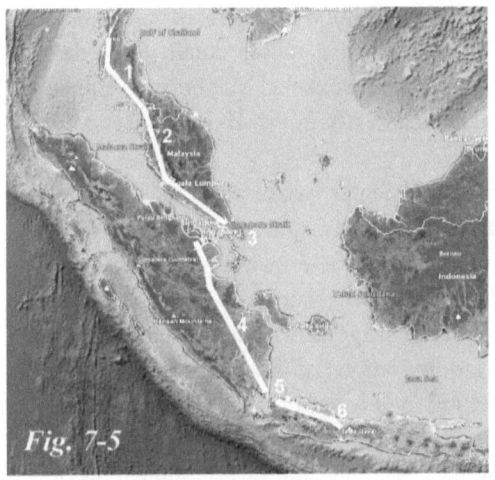

Fig. 7-5

Next came the coastline of what is now Malaysia (2), a distance of 712 miles. In southern Malaysia, they ran into a group of islands in the Singapore Strait (3). Once they were able to cross the Singapore Strait, possibly by hanging on to some jetsam or flotsam as one evo-illusion site suggested, they wound up on the immense island of Sumatra. I doubt these brave people were able to make boats so they could sail across.

So seafaring via flotsam and jetsam seems like as good of a theory as any for how they crossed. One site said these islands might have been joined by a mysterious but now non-existent land bridge. They had to travel southeast another 500 miles (4) to reach the south end of Sumatra, and what is now a 20-mile strait (5) that separates Java from Sumatra. They had to cross this strait; possibly again by grabbing on to some flotsam and, well, just floating across; or walking across on another illusionary land bridge. By now they had almost reached their goal. They needed to travel 356 additional miles to Java, where they finally rested, and took residence. Java must have been a terrific place, far better than the Africa they left behind, because these people remained in Java for the next, oh, 700,000 years. I guess they had had it with trekking. Any leaders who wanted to go trekking again were soon thrown out of office; or worse. What on earth would make ape-humans want to cross thousands of miles of land and sea in the first place? I can't imagine. What was wrong with simply staying where they were in Africa? The coastlines along all of the thousands of miles they trekked, with the exception of the deserts, would have provided excellent living environments. Why wouldn't they have stopped at those thousands of possible locations along the route that would have provided reasonably comfortable living? There was absolutely no reason for beings that had intelligence just above that of apes to wind up in Java.

After extensive digging on Java, paleontologists found two ape-like skullcaps, one human thighbone, and a tooth, to prove beyond a semi-shadow of a doubt African hominines made this incredible trek. The first two human migrations out of Africa were now complete. No bone remnants of either trek have been located along the route. But that's just a minor detail. Someday maybe someone will find some bones from the trekkers. Maybe not. Either way, this beyond amazing trek will remain steadfastly in evo-illusion lore. All I can say is, "Whew!" What a trip.

Remember, in the 1800's in the United States, when pioneers migrated halfway across the country to the Midwest, and a few all the way to the West Coast, to seek their fortunes? They did it with horse-drawn covered wagons. Their hardships were unbelievable. Many died. But many also made it. Well, the trip across the United States in covered wagons was less than peanuts compared to the difficulties posed by the migrations of Java man and Peking man. So I say kudos to Java man and Peking man for going where no man, well no sub-human, had ever gone before.

That pretty much takes care of the earliest migration of hominines out of Africa. After what I'm sure was careful consideration, the other hominines decided to remain in Africa for, oh, 640,000 more years. I guess they felt they had it pretty darn good right where they were. I mean, why travel naked and on bare feet all the way to Peking or Java, when you have so much fun and good stuff right there in Africa? Homo heidelbergensis must have made a similar incredible trek from central Africa to Germany, which was about as impossible as the treks to Java and Peking. I think you get the idea of the dauntingly impossible task migrating out of Africa really was, no matter what the end goal was for the hominines. Drawing migration arrows on maps is simple. Believing them is as well. For years I believed those maps with migration arrows without question. They seemed so logical. Why would anyone fake a story like that? But really considering the impossibility and ramifications of such tale is another thing entirely.[5-13]

Follow the Fur: Science makes numerous theories utilized to explain natural phenomenon. Then observations are made that either validate or disprove the theories. Scientists can then bend the observations and make up myths in an attempt to support faulty theories; or they can discard the theories and search for new ones. In the case of human evolution and human migration out of Africa, observations that disprove the theories hatch the fabrication of new myths and illusions so the theories can survive. Never do negative scientific observations cause the discarding of these theories. Human evolution and migration are never questioned. They are givens that must be supported, no matter what the evidence shows. Even the *thought* of discarding these theories is an anathema. The theories always remain in place, rock-solid, continually supported by new illusions. The theories must survive at all cost.

There are several very obvious markers that can be used for tracking and testing the validity of evo-illusion's story of human evolution and the migration of humans out of Africa. Did these really occur as evo-illusionists say they did? What do the most obvious markers have to say? One very obvious and telling marker is the tracking of *human skin*, versus the fur and skin of primates (apes and monkeys). If all humans and modern apes have an ape common ancestor, we should be able to utilize ape fur and skin markers, and human skin markers, to indicate whether evo-illusion's story of human evolution and migration is true. Ape fur is functional in all types of weather. Ape fur actually keeps ape bodies cool in very hot weather conditions. It also protects them from cold snaps. Apes can survive very cold temperatures because of their physiology and fur. Apes first appeared in equatorial Africa, where the weather is mild and pleasant, for the most part. It would seem they wouldn't really need heavily furred coverings. Humans that live in equatorial Africa can survive just fine without protective clothing. Why did apes evolve furred bodies where they didn't need it, whilst the humans that are their descendants evolved away all of that fur? If apes needed fur, why didn't humans? Or if humans didn't, why did apes?

One explanation concocted by renowned evo-illusionist Dr. Dan Leiberman is that early humans in Africa needed to run down their animal prey so they could kill and eat them. According to Dr. Leiberman, they evolved away their fur, and then invented and evolved sweat glands, which would keep them cool when they ran. Furred animals would become exhausted quickly, and collapse. Humans, without fur, and with sweat glands to cool them, would be able to outrun the poor beasts, catch up with them, and move in for the kill. But how often could this occurrence be significant? Could running down animal prey be such an evolutionarily significant event, and would it occur often enough, to completely change the outer covering of the entire human species over any period of time? Of course this is an absurd notion. But this is the highly accepted story given by Dr. Leiberman.[14]

Dr. Mark Pagel of the University of Reading in England and Dr. Walter Bodmer of the John Radcliffe Hospital in Oxford proposed a different but highly accepted solution to the mystery. Humans lost their fur, they say, to free themselves of external parasites that infest fur, like blood-sucking lice, fleas, ticks, and the diseases they spread. So, according to Pagel and Bodmer, apes evolved fur while they were inundated with lice, fleas, ticks, and the diseases they spread. They then kept their fur for millions of years, and still sport it today, despite the fact that they were and are still inundated with these bugs, whilst humans evolved away their fur to protect themselves from the evil bugs. I guess evolution pre-furred to protect 100% of humans, but not one of 625 primate species. One might think some primates would have evolved human-type skin so they too could be protected from the bugs. But that just isn't the case. Does this fit the theory of gradual random mutational changes over time? Nothing in the world of random can come out with a result of 100%; or 625 to 1. Not coin flips, not roulette wheels, and not random mutations. One hundred percent disproves evolution, so the story is bent so it will fit the theory.[15]

Two hundred thousand years ago Africa was populated by hundreds of heavily furred ape species, and one fur-less human species. The Human species that could survive in equatorial Africa unclothed, but not at all in cold climates, decided to venture north into Europe and Northern Asia... into very cold climates. Maybe they migrated in the summer months, when the weather was warm. Can you imagine their shock when the weather turned freezing? Their choice was to again evolve furred bodies. That would have been "easy evolution". After all, apes evolved fur in equatorial Africa where they certainly didn't need it. Evolving fur in the northern cold climates should have been a cinch for evolution. The heavier haired humans who could endure the cold would survive better than lightly haired humans. Fur should have returned in short order. But, alas, it didn't re-form at all. Humans remained fur-less for tens of thousands of years. There wasn't, and isn't, even a teensy sign of fur returning to humanity in any human group; not even the ones that live in subzero weather conditions. Fur, for humans, was finished.

Today there are no apes or monkeys, out of 625 species, *in the process of* eliminating fur and replacing it with human-type skin; not one. If evolution were valid science, this should be happening in numerous primate species. But it just isn't. Further, there are no apes in the process of evolving verbal communication, or toolmaking. Did they decide to do so several million years ago, then just quit? Did they get frustrated? Or did evolution from apes to humans not happen at all. The evidence says that's the case. Human evolution is just another evo-illusion.

Race Leaves Obvious Migration Markers: Another palpable tool that can be used for tracking and testing the validity of evo-illusion's story of the migration of humans out of Africa is by using racial characteristics as markers. Just like the human skin marker, this one is avoided like the Black Plague by evo-illusionists. It should be quite easy to compare and contrast evo-illusion's story of human migration out of Africa by simply following the distribution of racial types. Does it support the story of human evolution and migration? The story of human migration given by evo-illusionists, and those maps with arrows, are widely accepted but simply not possible.

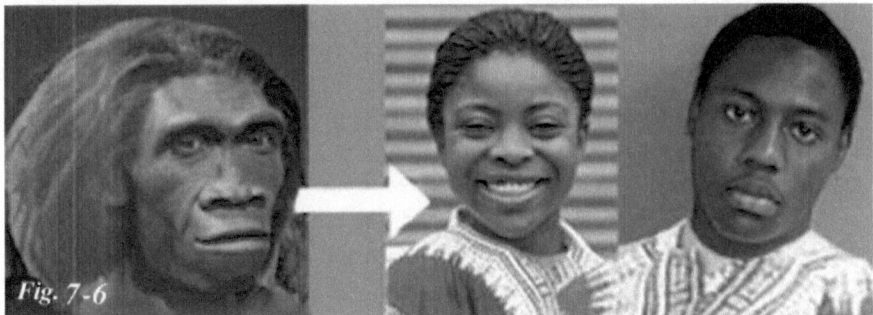

Fig. 7-6

The story of human migration allegedly begins with the earliest fossils of recognizably modern Homo sapiens that appear in the fossil record at Omo Kibish in Ethiopia, around 200,000 years ago.[6,7] According to the timeline and events proffered by evo-illusionists, Homo erectus hominines (left in Figure 7-6) that wisely remained in Africa evolved into modern-day black Africans (right in Figure 7-6). Black Africans populated the entire southern half of the African continent. There were no other notable races of humans, which is a very strange set of affairs for evo-illusionists. How could a random process produce such a pure result? Africa is composed of so many different environments: mountains, highlands, savannahs, grassy plains, deserts... One might wonder why a random process such as evolution would produce a single race of people rather than a varied group of races. Africa has varied environments, why wouldn't evolution produce varied races on that one immense continent? I'm certain evo-illusionists have that figured out. I just don't know what their solution is, so we'll leave it at that.

Africans didn't get the itch to leave Africa again until between 60,000 and 70,000 years ago. To evo-illusionists, the events that set this fourth and greatest migration in motion probably had something to do with major climatic changes probably driven by the onset of one of the nastiest segments of a major ice age. A cold snap probably caused a rapid decline in the black African population. Evo-illusionists say their population probably dropped to fewer than 10,000. According to evo-illusion, present-day humans depended on black Africans for their existence. They were our true ancestors. And they were probably holding on to their existence by the skin of their teeth. Probably.

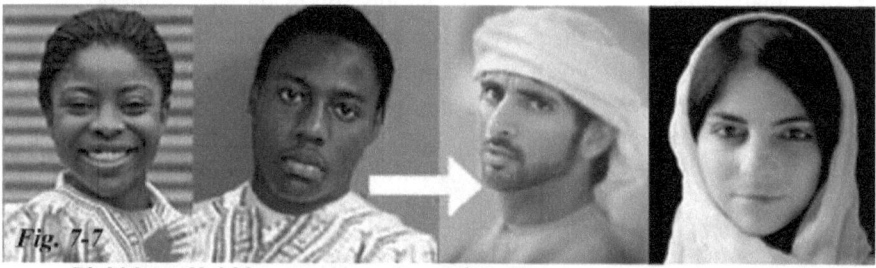
Fig. 7-7

70,000 to 60,000 years ago: Once the climate started to improve, black Africans made an incredible return from this near-extinction causing cold snap. Black Africans again completely dominated the southern half of Africa. According to paleo-illusionists, northern Black Africans did the amazing. In a very short time, they evolved into medium complexioned Middle Easterners (Figure 7-7). Evolution traded black African characteristics for Middle Eastern characteristics for some very strange reason. Northern Africa became dominated by Middle

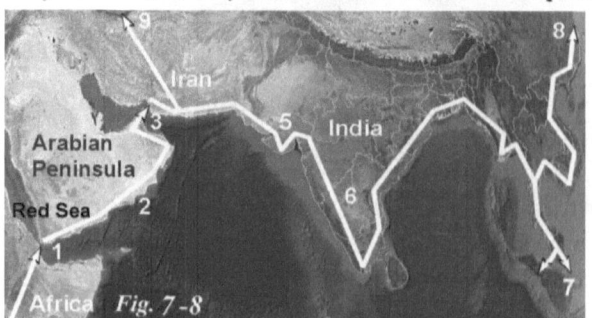
Fig. 7-8

Easterners. Middle Easterners also crossed the Red Sea and migrated to the Arabian Peninsula. According to evo-illusion, the earliest people to colonize the Eurasian landmass likely did so first by crossing the Sahara Desert; then by floating, possibly on flotsam, across the Mandeb Strait or the Red Sea (1 in Figure 7-8). The strait separates present-day Yemen from Djibouti. This is a 12 to 20 mile waterway, depending on where their departure and arrival points were. Crossing this straight is just another daunting task for the wanderers of 70,000 years ago. But somehow they made it across. These early beachcombers expanded rapidly along the Arabian Peninsula (2). They again risked their lives to cross the Strait of Hormuz (3). What kinds of boats could people 70,000 years ago assemble so they could traverse the Mandeb Strait and the Strait of Hormuz? Why would they do it in the first place? What kinds of resources were available for boat building anyway? But, let's assume they risked their lives, and successfully made the crossing because they had a huge, mysterious, inexplicable desire to do so. They took the same path across Iran, Pakistan, and India (5) that Java man and Peking man supposedly took hundreds of thousands of years earlier. Those medium complexioned Middle

Fig. 7-9

Easterners that travelled into India and Bangladesh evolved very dark-skinned Indian characteristics for some very strange reason. (Figure 7-9) Why would evolution make this change? Did Indian characteristics have an advantage over Middle Eastern characteristics?

50,000 years ago the trekkers reached Southeast Asia (7 in Figure 7-8): They crossed the Timor Sea to get to Australia, a distance of about 1,000 miles. Could the people in those days make a boat that would take them 1,000 miles in treacherous waters? If they left the safety of the islands around Java to venture to Australia in a feeble canoe, they had absolutely no idea Australia even existed. The women had to take the trip as well, or else Australia could not have been populated. The women that made this trip had to be incredibly tough. I can just imagine their discussion, when the adventurer told his spouse: "Get the kids and get in the canoe dear. We're going way way out to sea; way

way out there. I certainly hope there's some land out there that we can live on that's better than this stinking place. If there isn't land, well, we'll just have to die." She might have said: "If there isn't any land out there, I'm going to be soooo mad at you. You will never hear the end of it. How are you going to fix the leaks in the canoe? What about food? Did you pack water? If I told you once, I told you a thousand times… blah blah blah…" Oh boy, this was going to be a fun trip. I bet the guys never even stopped to get directions. In reality, any person that took this trip was idiotic, and soon dead; and the women would have been uber-right in thinking their uber-daring men were also uber-crazy.

Fig. 7-10

Groups of Indians migrated up the east coast of what is now China (8 in Figure 7-8) and evolved into the first Asians (Figure 7-10). Why did this occur? Evo-illusionists would have to say it was Indians that evolved into Asians because Indians were the only people available, unless Peking man evolved into all Asians. Middle Easterners populated only the Middle East at this time in the timeline. Middle Easterners that evolved into Indians left absolutely no modern trace of that evolution. There were virtually no Middle Easterners left in India after their remarkable evolution into Indians. Was there any difference in the survivability of Asians or Indians? I rather doubt it since both have survived so well. So what were the reasons for this change?

Next global warming took place. The ice age was over. Groups of medium complexioned Middle Easterners were able to migrate north to the Fertile Crescent and to the edge of the Black Sea (9 in Figure 7-8).

45,000 years ago: A mini-ice age occurred. Humans moved into what is now Turkey, Bulgaria, Hungary, and Austria. Medium complexioned Middle Easterners evolved into white complexioned Europeans (Figure 7-11), for some strange reason. Oddly, up until this point, every person on Earth had black hair. Europeans evolved red, brown, and blonde hair colors that had never previously existed. Again, why? Wasn't black hair color pretty darn good for everyone?

Fig. 7-11

45,000 to 40,000 years ago humans moved from southern and eastern Asia into northeast Asia. From Pakistan, they moved into central Asia. Japan became populated with Indians or Peking men that had previously evolved into Asians.

40,000 to 25,000 years ago Central Asians moved west towards Eastern Europe, and north into Eurasia, and above the Arctic Circle where it was dark for half of the year, and fur-eezing cold. Why would anyone leave the warmth of Africa for the freezing north? Why on Earth didn't they start a southward migration back to livable climates? They supposedly left Africa because of a cold snap. And they migrated to an area that was incredibly colder and dark?

25,000 to 22,000 years ago the ancestors of Native Americans crossed the Bering land bridge, what is now the Aleutian Islands. (Figure 7-12) I lived in the Aleutian Islands for two years. I was a dentist for the United States Navy stationed at the anti-submarine warfare base on Adak Island. The Aleutians being a land bridge from Russia to North America is unimaginable. Did that occur, and did people in Eastern Asia cross the land bridge and venture into North America, and later South America? Evo-illusionists say that during Ice Ages, so much water was sucked up into forming the vastly increased ice on the planet that the ocean levels dropped 300 feet forming the land bridge. But the

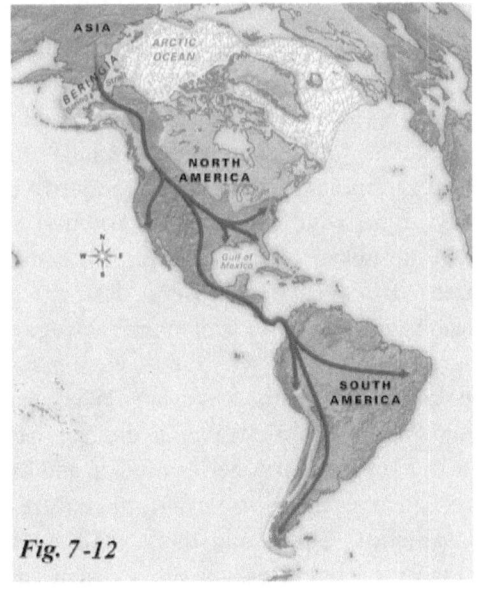

Fig. 7-12

waters around the Aleutian chain of islands are thousands of feet deep. Which means the ocean around the islands was still incredibly deep when the largest ice ages occurred. There was no land bridge that allowed trekkers to cross from Kamchatka Peninsula in Russia to mainland Alaska. Why would they even want to try? The entire area of northern Asia, the Aleutian Islands, and northern North America had to be freezing cold. Again, why weren't humans migrating toward the warm climates back in Africa and Southern Asia?

The land bridge notion is simply part of the evo-illusion needed to explain the impossible trekking and migration that supposedly took place. Every isolated island that has inhabitants forces paleo-illusionists to create a new land bridge. North America and South America are really incredibly isolated and immense islands that have inhabitants that must be accounted for in the "out of Africa" evo-illusion. So how did humans populate these isolated islands? Of course via the land bridge of the Aleutian Islands! Traversing the Aleutian Islands 22,000 years ago would have killed every person who tried. Plus, the distance between islands varies between 16 and 200 miles. In that freezing ocean with its incredible currents, and gawd-awful weather, the trip across the Aleutians in anything that boat builders of 22,000 years ago might have built would have been 100 percent fatal for all participants. This is just another impossible journey choice that somehow the early human trekkers perhaps accomplished. Maybe. Why would anyone want to take such a dangerous trip in the first place, when there was so much land and resources right where they were in Asia?

The big question is, who did the illusionary migrating into North America? Was it the Asian people of China? Northern Russians who resembled Northern Europeans? These are the only two possibilities, because these were the only people who inhabited the east coast of Asia at the time. Additionally, why did the trekkers evolve into brown complexioned American Indians and Mexicans? Did the evolution from Asian or Caucasian to American Indian and Mexican begin when they started their voyage across the island chain? Or when they arrived in Alaska? Why didn't these people remain as Asians, or as white Europeans or Northern Russians? What caused the racial change? Very puzzling. What is most puzzling is the fact that the North American population was nearly 100 percent American Indian and Mexican, two very similar groups. Before Columbus made his voyage, there were no white Europeans or Asians in North America. One would think some Asians and some white Europeans would have trekked across the non-existent land bridge of the Aleutian Islands as portrayed by evo-illusionists, and in the map above. The population of North America should have been made up of a mix of these groups. But North

America was racially pure Mexicans and American Indians, and completely devoid of Asians and Caucasians. The story given doesn't fit the racial markers at all. The only possible explanation is that 100 percent of Asians and Caucasians that crossed the Aleutians into North America almost instantly evolved into American Indians and Mexicans since there were zero Asian and Caucasian "leftovers" in North America until European explorers brought their hordes. Evolution, being a random process, should have left populations of the people in North America that made the crossing. North America should have been racially very mixed with Asians, and Caucasians. Is the racial makeup of North America before 1492 evidence that there was no crossing of the Aleutians whatsoever, nor migration into North America by Asians or Europeans?[5-13]

In 1804 Thomas Jefferson had congress appropriate $2,500 to pay for an expedition led by Lewis and Clark from the Mississippi River to the west coast of the United States. Their route was much farther south than that taken by the trekkers of 22,000 years ago, which would allow them much better weather conditions. They also had warm clothing, tents, horses, mules, canoes, plenty of water, guns to hunt, compasses, and maps. The Lewis and Clark expedition was an incredible undertaking that required months of preparation and training for the 29 members of the team. If they weren't well trained and well supplied, they would have all perished. One site noted:

The Lewis and Clark Expedition left from St. Louis, Missouri in late May 1804, proceeded up the Missouri River until its tributaries ended in the Rocky Mountains, crossed the Rockies and down the Columbia River to the Pacific Ocean. The trek was extremely arduous over 28 months and 8,000 miles on the trail. The Lewis and Clark Expedition's findings forced the nation to face the harsh reality that no easy all-water route to the Pacific existed.[16]

Is it believable that, 22,000 years ago, people who were barely able to start campfires and to break rocks to make tools could cross the North American continent from the west coast of Alaska to the area where New York is now? Most of the areas on the trek were not survivable for the humans of 22,000 years ago. But somehow they did it, and did survive. The arrows drawn on migration maps prove they did. Why didn't they travel south along the west coast of North America? They would have wound up in *Sunny Southern California*. Why did they pick the most difficult route imaginable? What's really strange is that American Indians used very feeble canoes to traverse waterways. Their canoes may have been great on rivers and lakes, but they wouldn't have been capable of crossing any oceans that needed to be crossed by their ancestors that migrated out of Africa. What happened to the ability of the ancestors of American Indians to make boats sturdy enough to cross treacherous straits and oceans? Didn't they pass this ability on to their offspring? The fact that American Indians didn't make boats large enough or sturdy enough to traverse seas and oceans is another indicator that kills the notion that humans populated the world from Africa. Maybe an immense case of dementia broke out among *all* North Americans, and they *all* forgot how to build seaworthy boats.

A far larger problem for evo-illusionists is explaining how Hawaii, Samoa, Fiji, Tahiti and all other large Pacific islands were populated. All of these islands, for the most part, are at least 2,500 miles from any other landmass or island, including each other. An exception is Fiji, which is about 630 miles from Samoa, still an incredible distance. There is no possible way a native of any of the major continents or large islands such as Australia or New Zealand could have made a raft or canoe and survived the journey to any of these islands. Adventurous humans that did leave any continent on any type of floating vessel would have had no idea there were islands thousands of miles away. As far as they knew, the whole world was made up of only the land they lived on. Their trips would have been completely blind, without any sort of guidance system. They had no maps, no compass, no knowledge about using the stars for navigation, and no idea about *any* potential destination. Try to imagine natives of Hawaii leaving those beautiful islands when they already had an abundance of land space, extraordinary food sources, and the population was sparse. Why would they leave? Hawaii has eight major islands. There was more than enough room for the natives of any ancient time. So what was the source of the native inhabitants of the Pacific islands? It's a much bigger puzzle than evo-illusionists want you to think. In fact, it's an immense puzzle. One paper said:

*The original inhabitants of Hawaii **are believed to** have come from the Marquesas Islands in the South Pacific about A.D. 250-450. Later migrations **probably came from** Tahiti. **It is believed that** the Tahitians had the skills to make the journey in both directions. **The knowledge of navigation** and the **personal self-confidence** that allowed these people to sail in relatively small deep-sea voyaging canoes over this huge distance are inspiring. One wonders how many ill-fated canoes missed the Hawaiian Islands and disappeared anonymously in the abyss of history and the ocean... On a relatively small* *double-hull canoe these explorers would have needed to carry their **water,** **food, seeds to plant for more food, tools, livestock and familiar animals (such** **as pigs, chickens, and dogs), plus sufficient childbearing women to reproduce** **themselves.**[16]*

Boy, that must have been a fun trip. Does the writer realize the Marquesas Islands are 2500 miles from Hawaii? In fact from anywhere? The natives built canoes capable of taking their water, food, seeds, tools, pigs, chickens, dogs, plus sufficient childbearing women to reproduce themselves? On their tiny canoes? Were they nuts? How many "are *believed's*" and *"probably's"* can be put into any writing before the writing itself becomes complete fiction? The natives "knowledge of navigation" would be useless without maps and pre-knowledge of what their destination was. The fact is the voyagers didn't have any "knowledge of navigation." The first voyagers had absolutely no idea or knowledge of any other landmasses or islands that might exist in any oceans. Without destination landmasses, navigation cannot exist. How did the natives get to the Marquesas in the first place, since the Marquesas are thousands of miles from any other island or continent? The writer's theory is easily testable. Have a group of uneducated Hawaiian natives make boats by hand, not with power tools, and see if they can make a return trip to the Marquesas. Actually, forget that idea. It would be a certain death sentence for all of them.

15,000 to 12,000 years ago the climate began another warming trend. The west coast of North America and the entire coastline of South America became populated. The immense question is why did South America become populated at all? North America provided everything any person could have possibly wanted and needed. Migration to South America certainly didn't occur because there wasn't enough land, food, or resources in North America. North America was sparsely populated, and the natural resources were overwhelming. The land was immense and free for the taking. Every kind of climate was available to those who may have wanted a change. There was absolutely no reason to make the immense migration to South America, particularly when no person on Earth knew it existed. So again, I ask why?

11,500 years ago Central North America and North East Canada became populated. People moved from the Bering-ean refuge and evolved into Eskimos, and Aleuts, the natives of the Aleutian Islands.

10,000 to 8,000 years ago the end of the Ice Age brought the beginning of agriculture. Amazingly, the Sahara got rid of all of its 100,000 cubic miles of sand and its gravelly mountainous wastelands, and became grassland and savannah again. That would sure be a natural wonder to watch. And then it reversed itself again, and became the desert it is today. Oh that Sahara desert! I wonder if all of the other deserts of the world go through cyclic changes like the Sahara.

It's very obvious that, analyzing the distribution of human racial characteristics, it can be shown beyond a shadow of a doubt that humans didn't migrate out of Africa and spread throughout the world as highly accepted *maps with arrows* indicate. Additionally, impassable geologic barriers, the deserts and oceans, would have stopped any migration cold. The idea that any humans or near humans would have even wanted to make such death-defying migrations is preposterous anyway. The story of human migration proffered by evo-illusionists simply makes no sense.

I wanted to make sure I didn't miss any pertinent new information that might challenge the information I've written so far. In writing this chapter, I researched numerous publications and documentaries on the subject of human migration. Public Broadcasting Service made a documentary on human migration titled *The Journey of Man: A Genetic Odyssey*. Of course PBS is very scientifically accurate. They wouldn't present a documentary filled with illusions... now would they? The star and moderator was a Ph.D. biologist named Spencer Wells. Wells earned a Bachelor of Science in Biology from the University of Texas at Austin in 1988, and a Ph.D. in biology from Harvard University in 1994. He was a postdoctoral fellow at Stanford University between 1994 and 1998, and a research fellow at the University of Oxford between 1999 and 2000. As of this writing he is a leading population geneticist and director of the Genographic Project for National Geographic. There couldn't possibly be a better source for the most updated information on the migration of mankind than Spencer Wells and this PBS documentary.

Wells research is centered around markers on the human Y chromosome that can only be passed from father to son. Coding on the Y chromosome passes mostly unchanged from generation to generation. According to Wells, these markers can be used as an excellent tool for tracking historic human migrations. The DNA that makes up human chromosomes is like a twisted zipper. When a copy error occurs, two of the "zipper teeth" nucleotides don't match, as they should. The result is they can't bond to each other. There's a gap in the DNA "zipper teeth", which makes it unique and traceable. Imagine if, due to a zipper machine error, two teeth on the zippers it made didn't touch or link up because they were too short. The clothing with the faulty zippers could be tracked all over the world. They and their delivery routes could easily be traced back to the zipper machine that made the faulty zippers.

The DNA markers Wells uses work very much the same way. He takes blood droplets from the fingers of his test subjects so he can isolate and map their DNA. Using this technique, he traced a particular marker called M-130 to a tiny tribe, the *San Bushmen*, in central Africa. My first question is, how was this tribe selected out of the over one billion people that populate Africa? Just think of the number of samples that had to be taken to eliminate all groups in Africa other than this tiny tribe. Wells goal is to prove that mankind migrated out of Africa to the southern coasts of the Middle East and Asia and into Indonesia and Australia. Supposedly his group of Bushmen has the most markers per population, so they must be the originators and common ancestors of every human on Earth that migrated out of Africa. We are all related to the San Bushman. Well, according to Wells anyway.

In the documentary, Wells himself tells the audience his feeling about the story of human migration:

Listen, I'll be honest with you. I've got problems. I've spent ten years checking and rechecking the details of this journey, until I have complete and total faith in our results. And the upshot? **The story, well frankly, it's impossible.** *If our ancestors made the journey I believe they did,* **they would have to be superhuman.** *The strength and resilience needed to conquer the world* **defies belief.** *And yet, there it is, written in our blood. What do you do when ten years of work leaves you with more questions than answers?*

Just as I did, Wells realizes the migration of humans as told by evo-illusionists is impossible. But he's going to prove it happened anyway. Why wouldn't he try to find another explanation for his markers instead of trying to prove an impossibility? Wouldn't good science do that?

In the documentary, Wells then flew to Africa, and met with the San Bushmen tribe, some of who supposedly migrated out of Africa 60,000 years ago. The tribe communicates with its own unique language. It's full of "clicks" and strange sounds not heard in any other language outside of Africa. Their very unusual language and sounds never spread to any other continents. Shouldn't this be an indicator that these tribesmen didn't migrate to any other continent? After all, the clicks and unusual sounds are still thriving 60,000 years later in the language of the San Bushmen. Their unusual language should be a huge marker, and make tracking migration routes a cinch. Language markers should be far more prominent, and a far greater indicator of the path of migration than a genetic marker. Language markers prove that this tribe didn't migrate at all. The excuse from Wells is, "The group that lost these clicks left, and then spread throughout the entire world." A notion so preposterous, it's not worth discussion. Why would only tribesmen who eliminated their clicks be the ones who migrated? So instead of following the very obvious plainly observable language markers, which disprove the great migration, Wells goes for DNA markers that only he and a few others can see; the invisible markers.

Fig. 7-13

Not only is language a marker, but racial and cultural characteristics are as well. No populations outside of Africa have the characteristic that these tribesmen have. The San Bushmen (Figure 7-13) are very uneducated and backward people who still hunt with spears. Do these people look like they could or would have ever cared to migrate to Australia? In the documentary, they are completely puzzled why Wells, this strange-looking foreigner, has any interest in them. Wells tries to explain DNA and the evolutionary tree of life, using a real tree for an example. The bush people are puzzled. The chance that this tribe, which is still barefoot, mostly naked, and uses spears to hunt their food, crossed the Sahara desert and many other treacherous land and water hazards approaches one in infinity. Even so, Wells is dazzled by the San Bushmen, almost in a worshipful way. It's as if he's talking to people who existed tens of thousands of years ago; and who are the common ancestors of all people who populated the entire Earth.

It would be quite easy to test evo-illusion's current "Out of Africa" theory by simply seeing if the San Bushmen would be able traverse the Sahara Desert, cross the Red Sea, traverse the Arabian Peninsula, then the Strait of Hormuz. They could use only their modern tools, spears, and boats; the stuff they have now, in the twenty-first century. After all, their technology is 60,000 years more up to date than were the implements used by the first San Bushmen trekkers. Actually, the test really doesn't need to be done. In the interest of saving the lives of the Bushmen, just using our imaginations is enough of a test.

They would have zero interest in doing the migration; and if they did, they would all die.

The documentary goes on to explain why our ancestors left Africa. The reason? Because between 72,000 and 50,000 years ago the seas receded forty kilometers from where they were due to a drop in temperature. The growth of the Arctic and Antarctic ice extracted an immense amount of water from the oceans. According to the documentary, savannahs turned into deserts, and the native population lost their access to ocean foods. It was an Ice Age and savannahs turned into deserts? But incredibly, the Sahara Desert turned into a savannah during the Ice Age so the tribe could cross. Does this story make sense? Did the central African savannahs turn into deserts, whilst the Sahara morphed onto a savannah? To fit Wells story and this documentary, that's what had to happen. In any case, Africa became a horrible place to live. So only the non-clicking San Bushmen made an impossible migration, one that was sure to kill them all. They trekked to places they had no idea existed, and that had far more horrible living conditions than Africa ever had.

The documentary goes on to explain how mankind almost became extinct due to the starvation caused by the ice age and lack of food. So what's a tribesman to do? Why, migrate across the Sahara, where there was no food or water; cross the Arabian Peninsula where there was no food or water, then traverse Iran, India, and Indonesia, and wind up in Australia! Wells states,

And no one knows how (they did it). Do you think that's impossible? I thought so too. But guess what. That's where our ancestors turn up next. **And no one knows how. There is no evidence of a journey by foot. And 6,000 miles of open ocean tells me that sailing is out of the question.** *So how did the descendants of the Bushmen get to Australia* **without leaving a trace?** *Maybe the answer is there.* (In Australia.) *Let's go find out.*

Is it a possibility that they didn't make this trek? Even during the Ice age there wasn't a land bridge from Asia to Australia. There were numerous spans of ocean to cross, not just one, that were hundreds of miles across. But that matters not. According to Wells, the Bushmen built canoes, and were able to paddle to unknown landmasses hundreds of miles into the ocean. The audience is even shown a small clip of natives launching wooden canoes, which is all that is needed to make them think these Bushmen could actually make this impossible voyage. Just a hint is all it takes, and the evo-illusion is in gear. The subject of canoe-travel to Australia is left for the imaginations of the viewers, just as all good illusions are.

Another great scientific test at this point in the documentary would be to have modern San Tribesmen make the crossing to Australia. Wells could see if the Bushmen could build boats capable of crossing hundreds of miles of ocean. He and PBS could have rescue crews following them so they won't die. Of course, they must begin their voyage without any knowledge of lands that might be their destination. Or here is even a better idea. Ask the Bushmen to give their best shot at making a boat. Let Wells see if he can reach Australia using the Bushmen's canoe. Think of how exciting this would have made the documentary. I guess no one from PBS thought of my great idea. Maybe next time.   Odds are Wells or the Bushmen would have gone around in circles endlessly, since that's what ocean currents tend to do, which would be very boring to watch. Plus, they would slowly die; which wouldn't be good for the documentary.

Wells says,

The most obvious route (for the tribesmen) would have been along the southern coast of Southern Asia. ***But so far there isn't one scrap of archeological evidence that they came this way. Neither has there been a trace of evidence in the genes of people living along the route. Something doesn't add up.***

Wells keeps repeating the same obvious conclusion that I came to when researching this book; that human migration as proposed by evo-illusionists doesn't "add up". There is no evidence it occurred. But Wells trucks on anyway. Wells flies to Australia, and with an archeo-geologist, they bend down and find stone tools! Of course these tools are very special, because after tens of thousands of years, they're still very sharp, just as if they had been made yesterday. Don't stone tools weather and wear from tens of thousands of years of erosion due to rain and blowing sand? In this illusion, they don't.

Wells and the geologist are astonished by black charred dirt that they find. (Figure 7-14) It was apparently singed by fire. Of course the black dirt is credited to the San Bushmen from tens of thousands of years ago. Don't

Fig. 7-14

these two realize that any singed dirt would have been long gone from rains and wind over those tens of thousands of years? In fact it would have been gone after the first rain. The geologist says the burned dirt is the remains of a small fireplace. Wells asks the geologist how old he thinks the burned dirt is. His answer? "Oh I'd say thirty to forty thousand years!" I'm astonished. How could anyone possibly come up with that number? Is this geologist super-human? Wells says, "Incredible. I mean it looks like it was done last week." The two men giggle. By millions to one, the greatest likelihood is that it *was* made by a group of Boy Scouts last week. Wells notion was right. Wells and the geologist walk over more of the area where the supposed Bushmen from Africa supposedly tread tens of thousands of years ago, with great reverence, as if they're walking on sacred ground.

Wells visits a group of Aborigines because he thinks they'll have clues about what became of the original tribesmen from Africa. He talks to the most educated Aborigine who says he is sure mankind started in Australia, and spread from there. Oops. No help here. He tries to explain DNA, but the Aborigine doesn't get it. So he's at a dead-end there in Australia. In the documentary, Wells doesn't check their DNA. Maybe he did off-camera.

Next he goes to India to look for genetic markers there. Wells collects hundreds of samples of blood in a small obscure town. DNA is separated from the blood in an effort to detect markers. Wells runs it through a machine that reads DNA coding. And *voila*! He finds just the copy error he needs; one single marker on the Y chromosome of one man (Figure 7-15). This proves beyond a shadow of a doubt that the tribesmen of Africa came through India on their way to Australia. Even though, as reiterated by Wells, there isn't a lick of physical evidence that this occurred. Wells immediately declares that the copy

Fig. 7-15

error in this one man's Y chromosome is 2,000 generations old. Somehow Wells has the same superpowers as did the geologist in Australia who declared the charred dirt was burned 30,000 to 40,000 years ago. Wells not only has superpowers, he is also has super-luck. Just think, out of the hundreds of samples he took, he finds one single error on one man's DNA, out of the more than one billion people in India. The man lives in a small village west of

Madurai. And this is all the evidence Wells needs to prove that mankind migrated from Africa to India, on their way to Australia. Wells says,

One microscopic change rewrites the history of an entire continent. And that feels pretty awesome. It proves that our ancestors did pass through here on their way to Australia.

It does? Does Wells actually believe what he's saying?

According to the documentary, in a very short time, humans proceeded with their migration from Africa, to India and Eastern China. But they didn't migrate to Europe until 35,000 years ago, ten thousand years after they migrated to Asia. The very brave people that inhabited Europe lived in caves where bears hibernated! Their skin was already getting more pale so it could absorb sunlight and get more vitamin D. How does Wells know this? The new Europeans wore clothing to keep themselves warm, so they had less skin area exposed to sunlight. They *needed* to be white-skinned. So why didn't mankind populate Europe before Asia? Well, they had to migrate around bad weather, which took them to Kazakhstan first. *Then* the people of Kazakhstan migrated north to Europe and east to Asia. How does Wells know? Because he found his DNA marker on one man from Kazakhstan named Nyasov. Which proves that this man's ancestors populated Russia, Europe, and Asia, and became Native Americans! He can tell all of this from just one tiny copy error on the man's Y chromosome. He completely ignores the racial markers that can be easily used to trace migration routes.

Wells drives all the way to Kazakhstan to meet this historic person, who of course, has no more idea how historic he is than does the Indian marker holder, or the San Bushmen in Africa, or the Aborigine in Australia. Wells goes to the man's house in Kazakhstan. In the photo (Figure 7-16), Wells is on the left. The historic Nyasov is on the right. Nyasov mumbles some stuff in Kazakhstani. Wells is worshipful, just like he was with the San Bushmen.

Fig. 7-16

After all, he is in the presence of a man whose ancestors he is sure produced all

Europeans, Russians, Asians, and Native Americans. He tries to explain DNA to the man, and about the X and Y chromosomes, and how his lineage takes us back 40,000 years. "So your Y chromosome that you got from your father...If we go back 2,000 generations..." Nyasov's wife is excited. He is told by Wells that his grandfather of two thousand generations ago was the "father" of all Russians, Europeans, and Native Americans. Nyasov giggles. Wells tells him that his Y chromosome has been here 40,000 years. The whole family giggles.

There is a tiny problem with Wells scenario. If this man's historic grandparents-of-two-thousand-generations-ago's offspring created a doubling in population, on average, only every 500 years, an absurdly and impossibly long time, for 40,000 years, starting with just his two grandparents-of-two-thousand-generations-ago, they would have produced 2.4 septillion...or 2.4 million million billion, or 2.4×10^{24} offspring. Remember, population doubling occurred in the 20th century about every 53 years, and during the Black Death in 450 years. If Nyasov's grandparents-of-two-thousand-generations-ago's offspring doubled in number only every thousand years, in other words, the original two grandparents-of-two-thousand-generations-ago would have, on average, only four descendants in one thousand years, and eight in two thousand years, and on and on, they would have two trillion, two hundred billion offspring today. (2,200,000,000,000). To have a rational number of descendants today, they would have had to double their family numbers on average only every 4,000 years. So what does this say about Nyasov being the offspring of grandparents-of-two-thousand-generations-ago that migrated to Khasikstan 40,000 years ago? It says Mr. Wells is wasting his life chasing down historically meaningless villagers. Wells should have calculated population doubling before starting his expedition. Lucky for me he didn't, because watching his journey is so entertaining.

Next Wells travels to northeast Russia to visit the *Chukchis* tribe. How on Earth he found and chose these unbelievably remote people to test I cannot imagine. On the first leg of his journey to the Chukchis, he takes a flight from Moscow to Siberia. It's unbelievably cold when he lands, which should be an indicator of the migration desires of the Siberian people who left Africa because it became cold, so they could inhabit uber-freezing Siberia. He tells the audience that it gets over 100^0F below zero here. Wouldn't that be enough to make these people migrate out of Siberia? My gawd. They live constantly in below-freezing temperatures, and their world is dark six months of the year, and that isn't enough to make them move? But Wells trucks on with his project to try to find someone with his DNA marker. This documentary just gets more and more astounding. He can't get any kind of flight to the Chukchis because of the unbelievable cold. He is stuck in a Siberian town for five days with the "hottest" day being -20^{0}F. It's interesting to note that Europe wasn't initially populated because of the cold, but Siberia was.

After waiting for better weather, Wells takes a helicopter ride to a village closer to the Chukchis. During the ride, he looks down from the helicopter and says, "How can anyone live down there?" When he lands he says, "Incredible. It's like being on the moon." Next he takes a bus ride 120 miles to another village where it's even colder. I wonder if the people who live here are concerned about global warming and carbon dioxide emissions. Hmm. Wells is so naïve, again, he has no idea he's killing his own theory. The more miserable the weather is, the worse it is for him and human migration. Cold weather was the stimulus for most migrations, according to Wells. The natives of Siberia haven't migrated south to warmer climes, away from their horrible weather, for tens of thousands of years. Very puzzling.

Wells next takes a six-hour ride in an old Soviet tank to his final destination. He finally meets up with the Chukchis. He again reiterates, "They spend their lives in temperatures that paralyze me." It's dark, foggy, and way below zero. The temperature dipped to -60^{0}F at night. Wells says he knows for certain that these Chukchis were able to cross the un-crossable Bering Strait 15,000 years ago, and into North America. Of course, maybe that magical land bridge that formed from Russia to Alaska allowed the Chukchis to simply walk across. From Alaska, in only 800 years, they populated all of North America and South America; and they evolved into American Indians and Mexicans. They finally hit the jackpot. All of those years of wandering aimlessly in sub-freezing conditions finally paid off. Again, in this documentary, Wells never did take a DNA sample of the Chukchis. Why? Did they not match his markers? Did he simply forget? Again,

Fig. 7-17

166

maybe he took the sample off camera and he simply didn't mention the results.

Wells next visited the Navajo Indians, since they are descendants of the Chukchis. Like the Aborigines, the Navajos think they came from where they now live. They think they were created in place; from the dirt. Wells tries to talk them into his notion about DNA markers, the phylogenetic tree, and that their ancestors came from Africa. He shows the Navahos a picture of Nyasov, and tells them they came from Nyasov's ancestors. The Navajos are interested, and kind of believe what Wells says. Again Wells doesn't do DNA testing on the Navajos; at least on camera. The importance of the DNA markers in this documentary fades. It ends with Wells in Rio de Janeiro, commiserating with Brazilians, where he doesn't do DNA testing either; not on camera, anyway.

Wells made another documentary for National Geographic's *Genographic Project*[8], which sports a diagram (Figure 7-17) that demonstrates the lineage of our human ancestors starting with two grandparents-of-two-thousand-generations-ago. In his own documentary he kills his entire life's work. He has no idea that he does. On his genealogy diagram, in only two generations, or in about forty years, there are eight descendants. Which means his population has doubled twice in forty years, an average doubling time of twenty years. In five generations, or about 100 years, there are twenty-two descendants. This means in 100 years, his original grandparents-of-two-thousand-generations-ago doubled their family size four and a half times. Feel free to count them yourself. Wells own chart shows average doubling occurred about every 22 years. The grandparents-of-two-thousand-generations-ago's ancestors in this chart represent the source of all North Americans, South Americans, and Asians. There were no "outside partners" to pair up with the family. But even if there were, the numbers don't improve for Wells. Average doubling times might increase to 100 years, which would still produce a near-infinite number of offspring. Of course, this chart is simply an example, and Wells would most likely say that's what it's meant to be and nothing more; that it's not an accurate population study. But it clearly demonstrates the notion of a 4,000-year average doubling time, which would be necessary for his 60,000-years-ago migration out of Africa to be valid, is completely impossible. So is his notion that humans came out of Africa 60,000 years ago. Wells notion is just one of many evo-illusions existent in the currently accepted history of mankind.

In a related article, Wells was asked,

If we all came from a black man, how did men and women of different colors come into being?

Wells answers,

The accepted explanation for skin color differences is that we first evolved in a tropical region, in Africa. The tropical Sun is quite strong, so the skin needed the protection provided by the natural sunscreen, melanin, which makes skin dark. When we started moving into the Northern Hemisphere 40,000 years ago, the Sun was not as strong. Anyone who's been to London in February can tell you that! And because the Sun helps us to synthesize Vitamin D, which we need to grow strong bones, we had to lose some of our pigmentation to allow enough sunlight through.[17]

Does Wells not realize that people who live in cold climates don't go around naked? They wear heavy clothing, which prevents the Sun from striking their skin. His answer is null. Further, Wells only discusses skin color in his response. Wells has severe tunnel vision. He ignores all of the other differences in racial characteristics that delineate black Africans, white Europeans, Asians, Indians, and Native American Indians. These characteristics would make excellent markers that could be utilized to track human migration. But for Wells experiment to work, he must ignore and do away with the most overwhelmingly obvious markers: language distinctions, racial characteristics,

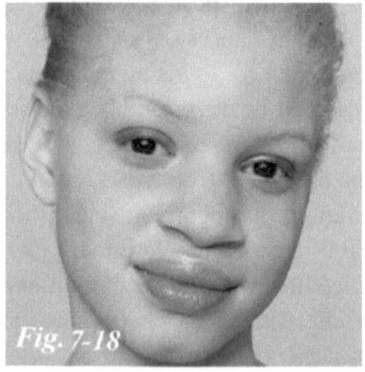

Fig. 7-18

and cultural idiosyncratic behaviors. He also must ignore the fact that he knows and reiterates constantly through this documentary that the migration of humans as posed by modern science is impossible, and nothing but an illusion. If black Africans migrated to Europe and evolved white skin due to their need for solar exposure, and resultant vitamin D synthesis, they would look like black Africans with white skin like the very pretty lady in Figure 7-18. She is a black albino; a black who was born without melanin. Black Africans who migrated to Europe wouldn't lose all of the characteristics that make them different from white Europeans and Asians besides skin color. There should also be black Africans in southern Europe where it doesn't get very cold. Europe should be made up of a mix of black Africans in the south and white Europeans that have black African characteristics in the north. But Europe was 100 percent Caucasian and Africa was 100 percent black before humans invented modes of long-distance transportation such as sailing ships. The illusion that blacks migrated north and lost their skin pigment so they could produce more vitamin D in cold climes doesn't compute. The overwhelming *visually* observable markers say Wells *invisible* DNA markers simply don't match reality. If black Africans migrated to Europe, and became white Europeans, the racial makeup of Europe would look far different than it does today.

The conclusions are easy in regard to the evolution of mankind in Africa, and the Great Migration out of Africa. They simply didn't happen the way evo-illusionists proclaim. The evolution of mankind in Africa, and the migration from Africa to all corners of the Earth would make much more sense if there were only one race and type of human. If a type of human that doesn't exist, *Homogenous Humans*, migrated out of Africa, and followed the migration paths given in virtually every anthropology textbook, it would make much more sense. But we humans are a grandly mixed race of people. We are made up of so many different types. Good science should utilize these types as markers in tracing the movement of ancient humans. The different races of humans say that the migration timelines and *maps with arrows* proffered by evo-illusionists are just what I say they are: illusions.

Population In Situ: So what is the explanation for the distribution of humans throughout the Earth and the geographic locations of the different races? There is only one plausible explanation: *Population In Situ.* It fits the evidence. The notion of *Population In Situ* is so absurd I hate to even write it. But here it is: It *appears* that the various groups of people on Earth first *materialized* in their own geographical location. Black Africans first appeared in Southern Africa. Middle Easterners in North Africa and the Arabian Peninsula. White Europeans originated in Europe, and Asians in Asia. North American Indians and Mexicans originated in North America. South Americans in South America. Tahitians in Tahiti, and Hawaiians in Hawaii. We humans didn't make magical impossible migrations that would be certain to kill fully 100 percent of all people who attempted them. Men and their "wives" didn't make canoes and fill them with water, foodstuffs, and livestock, and set out blindly in hopes of running into some other lands or islands in the vast oceans of the Earth. We originated where we first appeared. This notion eliminates the fable that we came from apes, and it eliminates the impossible migrations that supposedly brought us out of Africa. It explains the unique and exclusive geographic locations for each race of humans, and the nearly complete lack of geographic mixing of the different races. *And it makes absolutely no sense.*

But it makes far more sense than the story concocted by evo-illusionists. It comes from *objective and honest observation*. It fits the evidence. And it's a crazy notion. As ridiculous as this notion seems, it's far less ridiculous than the impossible journeys posed by modern evo-illusionists who make stories to fit their theories and illusions. The only scientific conclusions that can be drawn at this point in time is we humans have no idea why the universe is here instead of nothing, what human origins are, or how we came to populate the Earth. The more we discover, the farther we are from solving these great Puzzles. And, yes, those *maps with arrows* are illusions. The markers of racial characteristics, language, fur, skin, population clocks, the lack of hard evidence along the arrow routes, and our knowledge about human endurance says so. These markers trump any invisible markers Wells or any evo-illusionists may pose as valid.

I have now covered human evolution and migration. The next natural question that arises from these two discussions is, "If intelligent beings appeared here on Earth, is this something that could happen all over the universe?" This is a question that we all ask and think about at some time in our lives. The next two chapters are going to discuss what current science has to say. Of course some of my own new scientific insights will be added.

Chapter 8

The Universe Expands Exponentially in a Few Hundred Years

The search [for extra-terrestrial life] is a failure until that moment when it suddenly becomes a success. — Seth Shostak

Now that we have covered modern science's version of the beginnings of mankind and our migrations, the next natural question is, could this unbelievable scenario have taken place on other planets in the universe? Did cells form and come to life like they did on Earth? Did these cells stick together and form multicellular plants and animals? Did some animals form intelligence and consciousness? Did they migrate from a single landmass on these planets, cross massive oceans, and wind up inhabiting their entire land surface? Are there intelligent beings in the universe like us? The next two chapters will answer these and other fascinating questions to a near mathematical certainty. I don't want to come across as overconfident here, but when you are done reading chapters 8 and 9, you will have a pretty good idea about the chances of the existence of other intelligent beings in the universe. You will understand my confidence.

When I was a kid, I was enthralled, like most kids are, about the notion of alien beings on other planets. To this day I remember one of many movies I watched about aliens that had a profound effect on me. It was titled *Rocket Ship XM*. It was the

Fig.8 -1

story of four astronauts who took off for the moon in the Apollo 11 of the day. For this historic voyage, they were wearing street clothes! Rocket Ship XM took off successfully. But darn if XM didn't make an erroneous course correction. Instead of heading for the moon, it was headed straight for Mars.

Well, they thought they might as well land on Mars since there wasn't much else they could do. After all, it's tough to make a U-turn in space when you're speeding along at 25,000 miles per hour. So they went on to Mars, and made a perfect U-turn and tail landing. They got out to explore around, still dressed in their street clothes of course. They had small gas masks, with no oxygen tanks. Even as a kid I wondered about that. Next, they ran into Mean Martian Cave Men who started attacking them with rocks and boulders. They, of course, had to escape the Mean Martians by running back into their rocket ship for protection. They took off in the nick of time, and headed back to Earth. But, oops again, they were low on fuel. They didn't have enough to do their U-turn and tail landing. I was hoping beyond hope that they had an escape, but tragically there was none. They crashed into the Earth, killing all four on board. I remember thinking how brave they were, considering their plight. The movie ended with the senior scientist at mission control saying slowly and prophetically, "Yes, there will be another XM!" I remember those words as if it were yesterday. I already couldn't wait for the next installment of Rocket Ship XM. But as far as I know, and to my great disappointment, there was never a sequel. I was dazzled. I was so sad that they crashed, but my imagination was fueled. I thought it was the greatest movie I had ever seen.

I was so excited about the thought of alien beings, and spaceships to the moon and Mars. That movie really stoked an excitement in me that has never left. Like so many others, I'm dying to know if there are other intelligent civilizations in the universe. If so, what do they look like? Can we Earthlings ever know? Are they friendly? Could we communicate with them? I want so badly for these questions to be answered in a positive way. Will they ever be? In my first book, *Evo-illusion,* I discussed what I consider to be the *Puzzle.* I capitalize and italicize the word *Puzzle,* out of my great respect for what it represents. The questions I include in *the Puzzle* are so daunting, so beyond our ability to solve, the term deserves to be capitalized and italicized. The *Puzzle* is the ultimate enigma that has haunted intelligent scientists since the beginning of mankind. It has befuddled our best thinkers for so very long that they've found it necessary to make up fables in an attempt to form some kind of solution to the *Puzzle.* Amazingly, considering how far science has advanced, the fable makers are still active today. In *Evo-illusion* I included the questions of the origin of living cells and life itself in the *Puzzle.* I didn't include the question that also *needs* including: Is there extraterrestrial life? Is it intelligent?

Currently, there are two known *scientific* viewpoints that will determine whether or not there is or can be extraterrestrial life of any kind. They are mutually exclusive. Only one of the two viewpoints can be correct:

(1) Evolution: The theory that was made famous by Charles Darwin. He theorized that all of living nature was formed by *natural selection*. Decades after Darwin proposed his theory, genes were discovered. The theory changed to include *mutations*, or accidental copy errors in the genetic program that occur during procreation. Organisms and stages of development of biological systems became *naturally selected* if the mutational changes were beneficial. Those selected mutations brought about the evolution of all species and biological systems that exist today.

(2) IID: Some other entity or process exists that has an overriding and overwhelming intelligence and inventiveness, Intelligent Design, or as I term it, *Ingenious Invention and Design* (**IID**). **IID** was involved with the formation of all of living nature. I added "invention" to **ID** making it **IID** because invention is far more important in the process of the development of living nature than is "design". There are uncountable inventions in living nature.

One viewpoint of the two leads to the conclusion that life must be prevalent throughout the universe. Further, the chance of intelligent life on other planets in the universe is astronomically high. The notion that the universe was made only for life on Earth is absurd. There are over 100 billion stars in our galaxy, the Milky Way. There are over 100 billion galaxies like the Milky Way in the universe. It's likely that most stars have multiple planets, which means there must be somewhere around 10,000,000,000,000,000,000,000 (ten sextillion) possible locations in the universe that could harbor life. There are uncountable possible locations and chances for the existence of life other than on our home planet. There must be alien life, possibly even an enormous number of intelligent civilizations, throughout the universe.

The other viewpoint, on a purely scientific and mathematical basis, says no, there is absolutely zero chance of there being life on other planets. We are the only living organisms anywhere in the universe. The mathematical chances that life formed, even on one other planet, anywhere in the universe, borders on one in infinity. Life is so special, so unique. Earth is the only location in the universe with life and intelligent beings.

I will discuss the two viewpoints and how they affect the chance of there being life and intelligent civilizations other than our own in the universe in the next chapter. The existence of extraterrestrial life has profound ramifications for both scientific viewpoints. But first let's review the way mankind came to know the universe as we observe it today, with its near-infinite possibilities for the existence for extraterrestrial life.

Mankind has dreamed of the likelihood of living creatures inhabiting other worlds ever since we were able to determine what all of those twinkly lights in the night sky really were. Hundreds of years ago, many thought they were holes in the floor of heaven, and that we were looking at light originating from the other side of those holes. They thought they were peering into heaven itself. Then someone noticed that those twinkly little lights moved across the night sky, some at very different rates than others. The faster moving ones were the planets, of course. The slower moving ones were stars. Once telescopes were invented, the cat was out of the bag. As soon as early astronomers realized that we weren't looking at holes in the floor of heaven when looking at the night sky, the size of the universe grew exponentially in our mind's eye. Of course this expansion took place in man's *perception* of the universe. Since the universe exists only in our perception, (see Chapter 5 of *Evo-illusion*) it would be easy to say, as mankind's perception of the universe expanded, so did the universe itself.

Thousands of years ago, people thought the Earth was flat. What else could they think? If the Earth were a sphere, people and virtually everything that wasn't tied down on the bottom of the Earth would fall off. Everything fell *down*, so the same should be the case on the bottom half of the Earth. People couldn't travel very far, and the ball of the Earth is so large, it appeared to be flat. So the Earth remained flat in the minds of humanity for a very long time. Everybody lived on top. Next, enter the genius of Aristotle. Aristotle determined the Earth is a sphere using four different observations. He noted that everything falls straight down, no matter where you were on the Earth, as toward the center of a large sphere. When he traveled, he observed the position of the stars changed, which would not be the case if the Earth were flat. He noted that when ships sail away, the topmast eventually drops below the horizon, which means the Earth must be curved. The shadow of the Earth on the moon during an eclipse is always round. Putting these four observations together, Aristotle concluded the Earth must be a sphere. In 241 BC, Eratosthenes, a librarian in Alexandria, had measured the diameter of the Earth fairly accurately using the different lengths of the shadows of two sticks placed hundreds of miles apart, measured at the same time of day. Magellan's voyage around the world in 1543 AD established beyond a doubt that we live on a spherical Earth. The great puzzle was, why didn't people on the bottom of the Earth fall off? The answer became simple. The Earth, at that time, was the center of the universe, and everything fell toward that center. Puzzle solved. Science had advanced in great leaps.

But oops. Again, science was found to be incorrect. Nicolaus Copernicus published his theory in 1543 that positioned the Sun at the center of the universe, and not the Earth. (Figure 8-2) The Sun was motionless. The Earth and the other planets rotated around the Sun in circular paths at constant speeds. Since Copernicus calculated that the Sun was the center of the universe, thinkers of the day again had the problem of

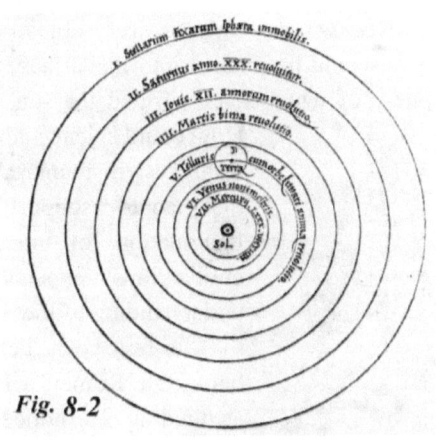

Fig. 8-2

trying to figure out why people and things on the bottom of the Earth didn't fall off since everything that falls goes down. Isaac Newton hadn't been born yet, and the people of Copernicus' time had no idea what gravity was. Copernicus must have fried his brain trying to figure out why planets traveled in circles around the sun. I know I sure would have.[1,2]

Then along came Galileo Galilei to straighten things out even more. I always wondered why Galileo went by his first name, and Copernicus went by his last. Anyway, Galileo had a small telescope, only about an inch and a half in diameter. Even so, he made some incredible astronomical observations with it. He saw vast clouds of stars in the Milky Way Galaxy, and realized the Earth wasn't covered by a dome of unmoving stars as the thinkers of the day had envisioned. He realized the stars and the night sky had tremendous, unimaginable depth. He observed the moons of Jupiter, which further evidenced the fact that the Earth is only one of several planets that circle the Sun and that also have moons. Galileo's excitement and imagination must have run wild. I bet he was dumbfounded with the thought that there were other planets circling the sun, with their own moons. He must have come up with the notion that there may be other civilizations on those planets. [3]

Re-enter the problem of the Earth having a top and bottom. Why didn't everything on the bottom half of the sphere fall off the Earth and into space? About a hundred years later along came Sir Isaac Newton to answer that question. In 1687 he introduced his theory of gravity in his book *Philosophiæ Naturalis Principia Mathematica*. The force of gravity attracted everything toward the *center of gravity* of every object. So no matter where you were on Earth, you would be pushed into its surface, toward its center. Gravity solved the mystery of why planets orbited the sun. Everything fit so perfectly now.

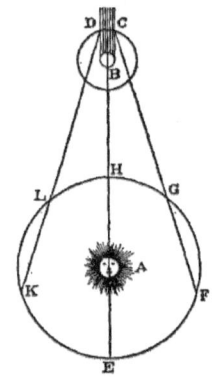

Fig. 8-3 Fᴵɢ. 70.

How could science advance any further? We now had a solar system mankind could understand, with planets that could support civilizations like ours. The imagination of mankind and the idea of alien civilizations expanded greatly with man's new understanding of the Sun and its planets.[4]

A few years before Isaac Newton published his book, Ole Romer, a Danish astronomer, really did the astounding. He noticed that the time between eclipses of *Io*, one of the moons of Jupiter, varied. Figure 8-3 is a reprint of Romer's drawing that helped him lay down the groundwork for the calculation of the speed of light.

Romer compared the apparent duration of Io's orbits as Earth moved towards Jupiter (F to G) and as Earth moved away from Jupiter (L to K). The time between Io's eclipses should have been constant; they were not. He charted the different times, and calculated that when the Earth was farthest from Jupiter in its orbit the eclipse cycle time was 22 ½ minutes longer than when the Earth was closest to Jupiter. He figured the greater eclipse cycle time must have been due to the light from the eclipse having to travel the diameter of the Earth's orbit around the

Fig. 8 -4

sun, about 187 million additional miles. The speed of light would then be 187 million miles divided by 1350 seconds (22½ minutes). This would give a result for the speed of light of about 139,000 miles per second; pretty close to current calculations of 186,000 miles per second. Figure 8-4 is a lithograph showing Romer busily at work.

Interestingly, he made all of the charts and diagrams that would have allowed him to calculate the speed of light. He never actually did the calculations; very strange, since all he had to do is divide 187 million miles by 1350 seconds. Maybe his computer was on the fritz. Many others determined a speed from his data, the first being Christiaan Huygens, a Dutch mathematician and a leading scientist of his time. Huygens work included early telescopic studies of the rings of Saturn and the discovery of its moon, Titan. For his work he constructed a twenty-three-foot long telescope that magnified objects about a hundred times, probably the most powerful telescope of the day. He also invented the pendulum clock, believe it or not. After corresponding with Romer and getting his data, Huygens deduced that light traveled 17 Earth diameters per second; or about 135,000 miles per second. So Romer and Huygens were pretty accurate considering their studies and calculations took place 400 years ago, using very rudimentary equipment.

Romer's measurements that lead to the first calculation of the speed of light gave rise to the *light-year*, a unit of distance measurement that astronomers use to determine and discuss the fantastic distances between the Earth, stars, and galaxies. A light-year is 6,000,000,000,000 (6 trillion) miles. Again, the invention of the light-year made it much easier to write about and deal with our new and expanding realization of the size of the universe. Think of having to constantly write and count those zeroes if miles were the unit used in astronomy. The light-year greatly simplified astronomical measurements. I am astounded at how brilliant these early astronomers really were. How could someone even think of charting the period of the eclipses of a moon circling Jupiter so he could determine the speed of light? And Romer did it with the comparatively weak telescopes of his day, which were not much more powerful than our binoculars. I will never cease to be amazed.[5]

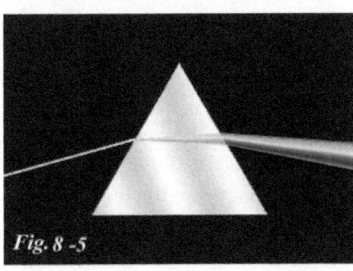
Fig. 8 -5

Around the time of the American Civil War, two British astronomers, William and Margaret Huggins were using spectroscopes to study the stars. A spectroscope allows a technician to separate the color of an object into its constituent colors. It works much like the way water vapor in the atmosphere separates light from the sun, which causes rainbows. Spectroscopes have a glass prism (Figure 8-5) that can separate a complex light beam into a band called a spectrum. The colors and their order in the spectrum of the object being observed are used as indicators for the elements present in the object giving off the light. Each element has its own spectral pattern, kind of like a human fingerprint has its own pattern. The spectrum of the Sun shows that it's made up of about 70% hydrogen and 28% helium. The Huggins focused their telescope on individual stars. They ran the light from those stars through an attached spectroscope and observed the resulting spectrum. To their astonishment, they found that all stars had the same basic chemical makeup as our sun. They are all made up of about 70% hydrogen and about 28% helium, which meant stars weren't tiny mysterious twinkly things in the night sky at all. They were all suns! And, our Sun is a star as well. They now knew that stars are so far away they just look like tiny twinkly objects. And humanity's perception of the universe again grew exponentially. Can you imagine what the Huggins must have been thinking when they realized the known universe is full of suns much like our own? Did they consider that each star, each sun, may have had its own solar system? Did they consider that each planet circling those suns may have had its own intelligent beings? I certainly would have, so they must have. Their excitement about their discovery must have been overwhelming.[6]

With each new discovery, the imaginations of scientists and those interested in astronomy must have run wild. Early telescopic images of Mars showed dark lines coursing back and forth over the surface. Italian astronomer Giovanni Schiparelli wrote a paper about those lines in 1877, calling them "canali". His paper sparked intense worldwide speculation about the probability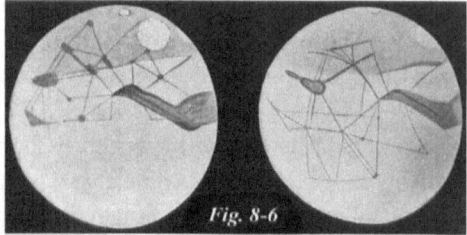

Fig. 8-6

of there being intelligent life on the red planet. American astronomer Percival Lowell exacerbated people's excitement and imaginations about the existence of Martians. He made intricate drawings (Figure 8-6) of Mars' surface markings, as he perceived them. Lowell published his views in three books: *Mars* (1895), *Mars and Its Canals* (1906), and *Mars As the Abode of Life* (1908). With these writings, Lowell, more than anyone else, popularized the long-held belief that these markings showed that Mars sustained intelligent life forms. I remember when I was a kid in the 1950's, looking at photographs of Mars with its dark lines. It was so exciting thinking about what those canals could be, and what kind of aliens made them. I remember another 1950's movie that had a great effect on me called *War of the Worlds*. It was about Martians that came here with an array of unstoppable weapons and un-damageable floating space ships. Their goal was to conquer Earth. Why, even an atomic bomb didn't put a scratch in their floating ships. Unfortunately for those Martians, our diseases, which they had no resistance to, finally killed them off. Movies like War of the Worlds stoked our wild imaginations about Martians, and canals, and conquering aliens. What a disappointment it was for me, and millions of imaginative astronomy fans, when probes sent to Mars that took photographs of its surface showed it was much like our moon: barren, inhospitable to life, and sterile. There were no canals; or Mean Martian Cave Men. The canals were just photographic illusions. Darn.[7,8]

Each new astronomical discovery brought with it a tremendous expansion of the universe in mankind's mind's eye. Scientific speculation about the possibility of life on other planets in the universe grew, and brought about more discoveries. One man was the king of all universe expanders. His name was Edwin Hubble. His discoveries expanded our perception of the size of the universe hundreds of billions of times, and allowed the realization that the number of possible sites that life and intelligent civilizations could exist is innumerable. Hubble is responsible for the fact I cite above that there are possibly 10,000,000,000,000,000,000,000 (ten sextillion) sites in the universe that could harbor living organisms and intelligent civilizations.

Fig. 8-7

Edwin Hubble's arrival at Mount Wilson, California in 1919 roughly coincided with the completion of the 100-inch Hooker telescope there, then the world's largest. At that time, the prevailing view of the cosmos was that the universe consisted entirely of the Milky Way Galaxy. When astronomers looked at the night sky through their existing telescopes they saw, mixed in with the stars and planets, uncountable tiny spirals of various shapes and sizes. These were called *spiral nebulae*. (Figure 8-7) Astronomers had no idea what spiral nebulae were. They were thought to be part of the Milky Way. Using the new telescope, Hubble took photos and compared the varying degrees of luminosity among *Cepheid variable stars*. Cepheid variables give off incredible amounts of energy, and so are easier to spot and measure. His comparisons allowed him to realize that one of the Cepheid variables was actually located in a spiral nebula. He realized that each nebula was, in truth, a star system like the Milky Way! All nebulae were much too distant to be part of the Milky Way and were, in fact, entire galaxies outside of our own. Each spiral nebula was actually a Milky Way, and there were hundreds of billions of them! Astronomers later determined the Andromeda Nebula, the closest major galaxy to Earth, was over 2.2 million light-years away. Ergo, it had to be its own star city, since the Milky Way is 100,000 light-years in diameter. All "nebula" were renamed "galaxy".[8]

Hubble's findings fundamentally changed the scientific view of the universe. Instead of the universe being made up of one galaxy, the Milky Way, composed of over a hundred of billion stars, it was now made up of hundreds of billions of galaxies, each one being composed of billions of stars. This discovery lead to the Big Bang theory, which had been proposed by Georges Lemaître in 1927. I can't imagine what it must have been like for Edwin Hubble, when he discovered what those tiny spirals really were. Just picture what went through his mind when he realized the universe was hundreds of billions of times larger than virtually every astronomer in the world thought it was. For that moment in time, he was the only person on Earth who knew the immense size of the universe. It's so interesting to study how so many astronomers came up with such astounding finds. I always wonder what it did to their psyche. I always try to imagine what it must be like to be the discoverer of such fantastic scientific upheavals. Considering the year of the discovery, I wonder if he went home, incredibly excited, and opened the door, and hollered out to his wife, as loud as he could, "Honey, I'm home...and guess what! The universe is so much bigger than I thought it was!" What a discussion they must have had at the dinner table that night! [9]

Once Hubble determined how immense the universe really is, it opened the door for greater speculation about the existence of intelligent life on other planets. Is the universe large enough to provide enough chances for intelligent life to form as it did on Earth? The next chapter will answer that question as fully as it can be answered.

Chapter 9

So Steve, Are Aliens an Illusion Too?

Intelligent life on other planets? I'm not even sure there is on earth!
— Albert Einstein

Now that we Earthlings have an idea about how immense the universe really is, and what it's made of, it's only natural for us to wonder if we're alone in the universe. It's really only been a few hundred years since we've been able to have that thought. Before we sent astronauts to the moon, and began sending probes to our various planetary neighbors, our hopes of finding otherworldly life were high. There were numerous locations in our own solar system that were likely candidates for life; possibly even intelligent life. What existed under that unbelievably thick cloud layer on Venus? Did those clouds cool the surface of the planet? Were there Venusians under those clouds? Mars looked like such a good candidate for intelligent life. After all, somebody had to be the maker of those giant canals. Sadly, the more we learned from our probes, rovers, and satellites, the farther we were from finding extraterrestrial life. The search for extraterrestrial life became far more discouraging. We earthlings have been reduced to grasping at alien straws; and we are spending billions of dollars doing so. The results have been so disappointing. I had high hopes that we would at least find unicellular life somewhere. Think how exciting it would be if we had found bacteria all over Mars, or on Venus; or anywhere. But as far as we Earthlings know, our solar system is sterile, with the exception of our beautiful blue planet.

An entirely new field of science has been organized and dedicated to searching for extraterrestrial life called *astrobiology*. Of course, the scientists that run the astrobiology show are called *astrobiologists*. I can't imagine actually being an astrobiologist, since there are absolutely zero indicators of life existing on any planet or moon that can be detected from Earth. I hate to say it, but what could be a more useless title than astrobiologist? There is absolutely zero astrobiology to study. Numerous television documentaries on this subject that I've watched always have several expert astrobiologists giving information about their speculations. "There may be life here, and it might be over there." To date it's neither here nor there. The organization leading the search for

extraterrestrial life is the *SETI Institute.* SETI stands for Search for Extraterrestrial Intelligence. They've developed a specialized radio telescope array (Figure 9-1) made to receive alien radio broadcasts. The hope is that there are intelligent beings on some planet beyond our solar system that are anxious to communicate with us Earthlings. The SETI group has constructed an array of 42 dishes at the Hat Creek Radio Observatory in rural northern California. When completed, the full array is planned to consist of 350 or more dish antennae, each 20 feet in diameter. The first group of antennae became operational in 2007. Completion of the full 350-dish array will depend on funding and the technical results from the first 42 dishes. If nothing is found with the 42, then the project will be stymied, and that will be that. From 2007 until 2015, the array has identified hundreds of millions of technological signals. So far, all these signals have been assigned the status of noise or radio frequency interference because they appear to be generated by satellites or Earth-based transmitters. If intelligent aliens were trying to contact us, they would have sent a constant flow of signals, and none of the unidentifiable signals received by the array have lasted long enough to qualify as being alien based. In other words SETI, as of this writing, is a big bust.

When a project such as SETI is a bust, there are two choices. One is to shut it down, and figure the odds of aliens on a nearby planet sending radio waves to us approach one in infinity. The other is to make up fantastical new scenarios that keep the illusion alive with congress, NASA, the taxpayers and donors that pay for such quixotic projects, so this futile search can continue. And that is just what happened. The new and exciting notion is that intelligent aliens may not be sending radio waves at all. They may have invented laser beams, and they may have constructed a gigantic laser light source, which they have beamed to Earth. If we could only find that laser beam, why it would be proof certain that aliens are trying to contact us. So now Jeff Marcy, a University of California at Berkeley astronomer who has found more exoplanets than any other astronomer, to his great credit, is now trolling for alien laser light beams, which eliminates some of that credit. Laser beams are made up of one pure wavelength of light. They don't spread out into a rainbow of colors when sent through a prism, like normal light that comes from stars. If an exoplanet did evolve life, and if that life became intelligent, and if that intelligent life invented lasers, and if that intelligent life found us, and if they directed their lasers at us to try to let us know they are there, well, we could detect that laser. We would know they are there. And... we could send them a message, saying we got their laser beam, and, and... then what? The notion intelligent beings on a planet orbiting a star close enough to us so that they could detect us, and that they are trying to communicate with us, is nothing but pure illusion. Which actually makes me very disappointed. Because there is no person who would be more excited than I if we could somehow detect life on other planets; particularly intelligent life. But that whole notion is based on an illusion that costs billions of taxpayer dollars to fund. In the rest of this chapter I will tell you why it's an illusion. Finding alien life just isn't in the cards for us humans. [1-4]

The first task to accomplish when thinking about and analyzing the chance that there are alien organisms on other planets and moons is to think backward: What would and should a planet look like if it were to eventually produce living organisms? Well, it must be assumed that all living things need lots of water and probably an abundance of carbon atoms. It's unlikely that life could exist based on any other chemicals, since on Earth there are millions of different species, and every one is based on water and carbon atoms. There are lots of other chemical choices here on Earth, but the only choice life has taken is carbon and water. The planet must be in what is called the "habitable zone", which means it's orbiting its star at just the right distance so that it won't burn off or freeze water and living organisms trying to catch a foothold on the planet. So the planet must be wet and warm. But above all, the planet must be able to somehow form biochemicals, or at least the organic chemicals that are needed to form living organisms. Since amino acids form proteins, there must be a huge number of ponds and lakes that harbor densely dissolved amino acids. The planet must be able to manufacture lipids from non-living sources. And there must be sugars like ribose sugar; the kind that makes up DNA and RNA. And there must be bases, such as adenine, guanine, cytosine, and thymine, as well; the bases that also make up DNA and RNA. Living cells manufacture all of these biochemicals, so how did they originate on a sterile planet? That question must be ignored if there is to be any discussion about the formation and origin of life. So, minimally, there must be ponds and lakes and oceans just chock full of these pre-biotic chemicals that make up life. The planet's bodies of water must be highly saturated with these chemicals. If that's not the case, currents and the natural stirring of the water would be constantly swirling dissolved chemicals, which would make it impossible for them to link-up in any fashion that would someday produce life. After all, chemicals can't just smartly swim toward each other and bond in hopes that someday they will form living organisms. So this is what a knowledgeable person would expect a planet in the throes of producing life to look like. What is really strange, the early Earth doesn't even look like a planet that one would expect life to arise. If we could take a walk on the sterile early Earth, it would be very difficult to predict that it would eventually be the home to billions of living organisms. Even today's Earth doesn't have bodies of water chock full of the biochemicals of life. They simply don't exist, which means they didn't exist billions of years ago when life took hold. It's highly unlikely that a planet exists anywhere in the universe in a habitable zone, loaded with bodies of water that are chock full of the pre-biotic chemicals of life. Thinking backward about the possibility of life on planets outside of the solar system isn't encouraging.

According to Seth Shostak, senior astronomer at the SETI Institute, there are six locations in our solar system that have astrobiologist fired up and hopeful that some kind of living organism might be found. They've long ago

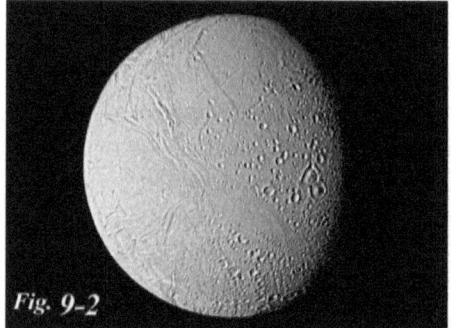

Fig. 9-2

given up the notion of finding intelligent life, so they're reduced to finding bacteria or the like. They are:

Enceladus (Figure 9-2) is one of Saturn's moons. Enceladus is one of the brightest objects in our solar system because it's covered by water snow. The snow reflects almost 100 percent of the sunlight that strikes it. In 2005, NASA's Cassini spacecraft photographed geysers of water spewing from cracks in Enceladus' southern hemisphere. Scientists think reservoirs of liquid water lie beneath the frozen surface that are warmed by gravitational interactions between Enceladus and Saturn. Some astrobiologists think the necessities for life are there, and maybe *Enceladans* are as well.

Of course this is nothing but a scientific pipe dream. Water isn't the only "necessity for life". A "few" other major ingredients are necessary, and Enceladus doesn't have any of them. It's a tiny moon. It has a diameter of 313 miles. Because Enceladus reflects so much sunlight, the surface temperature is only -330°F. The Rocket Ship XM astronauts, wandering around Enceladus in their street clothes, would be frozen solid in seconds. So wouldn't any random attempt at the formation of living cells do the same?[5]

Mars is the most popular site for astrobiologists hunting for otherworldly life. The dark stripes that appear in the Martian summertime at Horowitz crater (Figure 9-3) are particularly fascinating. Astrobiologists hope these lines are from salty meltwater. They're hoping to find indicators of life only a short distance beneath Mars' dusty outer surface. One astrobiologist stated that a "relatively simple probe" could sample this

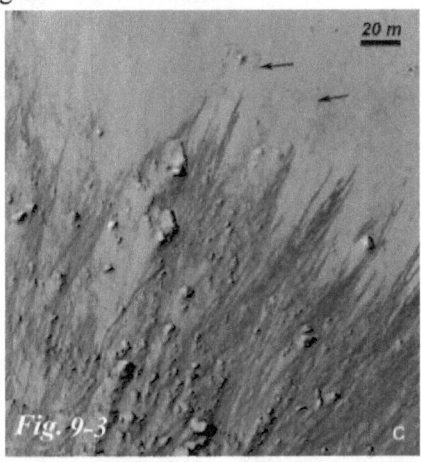

Fig. 9-3

muddy environment. There really is no such thing as a "relatively simple probe" when talking about Mars. The trip to Mars, and the rendezvous, is fraught with hazards and potential failure. As of this writing, forty-four attempts have been made at sending probes to Mars. Twenty-five of those have failed. The images sent back by the twenty-one probes that successfully made the trip were nothing but breathtaking. The search for signs of living organisms has been nothing but disappointing. The first two successful landers, Viking 1 and 2, were sent to Mars in 1976. I remember being very excited about the possibility that the landers might discover signs of life. I was almost sure they would. Vikings 1 and 2 were designed to run experiments that might find evidence of microscopic life in the Martian soil. Each experiment was performed multiple times on different soil samples. They provided conclusive evidence that Mars was sterile. NASA reported that the results were inconclusive, but that was, of course, to keep up the hopes of the American taxpayer and congress so the money pump would remain primed. Mars has been found to be sterile by all of our landers, but if that were announced, why would there be a need to continue spending billions of dollars for a futile search? But sterile results sure didn't stop the search. Many more landers were sent since Viking. They had the exact same result. All showed clearly that Mars is hopelessly sterile. Mars was our greatest hope for finding life, and so far it's been a complete bust. Because the atmosphere is so thin, water can't exist in liquid form on the surface; probably not a few feet deep either. Water would act like dry ice does on Earth. It would turn directly into vapor from ice, which makes the possibility of life on Mars less than miniscule. Also, the notion that liquid water existed beneath the surface leaves out the possibility that photosynthesis or any similar type of energy gathering biochemical cycle evolved on Mars since it wouldn't be exposed to the sun. Mars is a little more than half the size of Earth. It has a diameter of 4,212 miles. The temperature on Mars may reach a high of about 70^0 F at noon, at the equator in the summer. Hey, maybe those astronauts in *Rocket Ship XM* could have run around in their street clothes after all! The low temperature on Mars is about -225^0 F at the poles. If the XM astronauts went to the poles in their street clothes, they would be frozen solid in seconds.

The notion of sending astronauts to Mars is nothing but absurd. The costs would make the funds NASA spent sending astronauts to the moon look like chump change. Mars is

Fig. 9-4

nothing but a giant sterile vacant lot. (Figure 9-4) Not worth spending trillions of more dollars on; certainly not worth killing a crew of astronauts for, which would be a likely result. Even sending more satellites and landers to Mars is a tremendous waste of taxpayer dollars. We know just about everything we can possibly know about Mars. Scientists have mapped its entire surface; they know what it's composed of. Every experiment they've run on Mars shows that it's sterile. NASA thinks if they find water on Mars, well, it may have life; or it may have *had* life. The existence of water does not mean life; it's billions of steps from life. It's time to say good-bye to Mars. But NASA will continue sending probes, and making people, AKA the American taxpayers and congress, hopeful that there may be life over there, or here, or maybe at the poles… If taxpayers are kept hopeful, no one will bitch. NASA can keep

Fig. 9-5

spending huge amounts of money on projects that their scientists already know what the result will be: no life will be found.[6-9]

Titan (Figure 9-5) is Saturn's largest moon, and the only body in the solar system (besides Earth) known to have liquid lakes. Unfortunately for astrobiologists, these aren't liquid *water* lakes. They're made up of ethane and methane, which are natural gasses on Earth. A constant rain of hydrocarbons keeps the lakes filled. Titan looks a great deal like the planet Venus. It has a completely cloud-covered surface that cannot be seen from space. It's surrounded by an orange haze that kept its surface a mystery for Earth's scientists until we sent landers.

Titan's atmosphere is active and complex, and is composed of 95% nitrogen and 5% methane; not a very hospitable atmosphere for living organisms. NASA, the European Space Agency and the Italian Space Agency sent *Cassini*, a combination satellite and lander that orbited Saturn in 2004. The lander touched down on Titan in 2005. Cassini's view of Titan is a composite photo, taken using a filter sensitive to infrared light. It's able to "look" through clouds, and show the huge lakes on Titan. Some are as large as Lake Superior on Earth. Despite the odd ingredients and Titan's average temperature of -290⁰ F, astrobiologists have high hopes that life could have formed there. As far as is known, there is no water, which makes Titan a sterile ball, and a less than unlikely place to find life of any kind. Titan has a diameter of 3,200 miles. Even though its name gives the impression that it's the largest moon in the solar system, it is not. Ganymede, a moon of Jupiter, is the largest, with a diameter of 3,273 miles. If Titan or Ganymede were to break free from their planet's gravitational pull, they're large enough to be classified as planets. For reference, both are larger than the planet Mercury, which is 3,032 miles in diameter, and our Moon, which is 2,159 miles in diameter. To give you an idea of their relative sizes, if the Sun were an eight-foot ball, Earth would be the size of a nickel and Titan would be the size of a pea. Interestingly, Ganymede, the largest moon in the solar system, is devoid of an atmosphere and surface liquid. The *Rocket Ship XM* astronauts, if they goofed up their course correction and wound up on Titan instead of Mars, and tried exploring around in their street clothing, would be frozen solid in seconds. [10-12]

Europa (Figure 9-6) is probably the second most interesting place in the solar system to look for life for astrobiologists, since there's probably more liquid water there than in all of Earth's oceans. The downside is that Europa's vast, salty seas lie

Fig. 9-6

beneath about ten miles of ice. Not only would it be difficult or impossible to get a probe beneath Europa's icy armor, but its oceans, if they are present, are pitch dark. This means photosynthesis is not a possible energy supplier for life, if it does exist. Ever-optimistic astrobiologists are hopeful that *something down there* may subsist on "geothermal heat or complex molecules…" Hey, maybe we can send some astronauts to cut a ten-mile deep hole, and then slide down to the seas, and go diving underneath the ice to see if there might be bacteria down there. Some optimistic astrobiologists think that beneath Europa's surface, active volcanoes may also heat the water, providing vents where bacterial life may thrive as it does on Earth. The keyword here is "may". "May" allows for anything. NASA's Hubble Space Telescope discovered an area where astrobiologists could search for the telltale signs of life: geysers of water vapor erupting from Europa's southern hemisphere. The optimistic astrobiologists want NASA to send a flyby probe to sample Europa's seas from a distance. Bill Nye, the science guy said, "The seawater is spewing into space." He urged scientists to fly by and "look at what's collecting on the windshield." And of course all taxpayers should be happy to pay for this great idea. Europa is primarily composed of silicate rock, and probably has an iron core. It has a diameter of 1940 miles. It's slightly smaller than our Moon. Its surface temperature at the equator never rises above -260^0 F. At the poles, the temperature never rises above -370^0 F. It's unimaginably cold, way too cold for the *Rocket Ship XM* astronauts in their street clothes. They would all be frozen statues in seconds. So would any attempt at forming life.[13-15]

Venus is another in the line of possible sites that SETI claims might harbor living cells. Figure 9-7 is a composite radar photo of Venus taken by the Magellan spacecraft. Venus and Earth are often called twins because they're both rocky planets and very comparable in size. However, the similarities end there. Venus takes 243 Earth days to spin once on its axis. Just think, night

Fig. 9-7

would be 122 Earth days long! Strangely, Venus rotates on its axis in the opposite direction than all other planets. On Venus, the Sun would appear to rise in the west and set in the east. (On Earth, the Sun rises in the east and sets in the west.) The surface of Venus is a scorching hot 850^{0}F; hot enough to melt lead. Venus is the hottest planet in the solar system; even hotter than Mercury, which is the closest planet to the sun. Venus' thick atmosphere traps heat. The atmospheric pressure is 90 times that of Earth's. *Rocket Ship XM*'s astronauts, dressed in their street clothes, would be both crushed by the atmosphere and evaporated in seconds. Good thing XM's course correction error didn't take them to Venus! The planet is generally assumed to be as sterile as the Sun on a hot day. But ever-optimistic planetary scientist David Grinspoon, astrobiology curator at the Denver Museum of Nature and Science, points out that high in the Venusian atmosphere temperatures are "refreshingly tolerable". Refreshingly? He thinks atmospheric sulfur dioxide and carbon monoxide might serve as food for floating microbes. Hey, this might be a great place to vacation sometime in the future. Plus

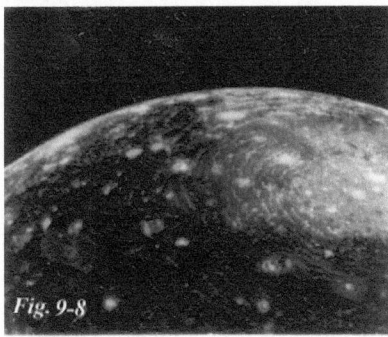

Fig. 9-8

visitors would get all of the sulfur dioxide and carbon monoxide they can eat. Venus is 7,521 miles in diameter. [16,17]

Callisto and Ganymede: The last of the "likely six" locations in the solar system that might harbor life are two moons of Jupiter, Callisto (Figure 9-8) and Ganymede (Figure 9-9), according to Seth Shostak. He considers these two moons of Jupiter together, and thinks they're "neck-and-neck candidates for biology". Like their more celebrated

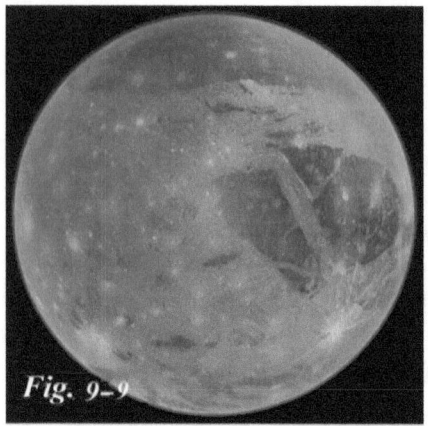

Fig. 9-9

neighbor Europa, Ganymede and Callisto may have buried liquid oceans. However, in the case of these two satellite siblings, briny oceans would underlie at least 60 miles of rock. If this is the case, why even speculate? Humans will never know if briny seas exist 60 miles below their surfaces. We can't drill 60 miles deep on Earth, we certainly never will on Callisto or Ganymede. The earliest life on Earth was thought to utilize photosynthesis to gain its energy from the sun. Sixty miles below the surface of either of these two candidates for life, there would be no sun, no photosynthesis, and no life. Callisto has a diameter of more than 2,985 miles; Ganymede's diameter is 3,270 miles. Daytime temperatures on the surface of Ganymede are in the range of -171^{0}F to -297^{0}F. The mean surface temperature of Callisto is -218^{0}F. Which means our Rocket Ship XM astronauts, in their street clothes, would freeze instantly on either moon. So would any attempt at life forming. Do Callisto and Ganymede look like they might be chock full of life? [18]

Of the six most likely locations for life in the Solar System, not one of them appears to have the slimmest of chances, except to hopeful astrobiologists. They all look like sterile spheres. I would like to make a major announcement: Other than our Earth, the entire solar system is sterile. The only planet that had a slight chance was Mars. But after sending numerous probes, every one of which showed Mars to be sterile, our ever-hopeful astrobiologists and scientists will continue their fruitless tasks. More probes will be sent, and all will turn up nothing, except new pictures of an immense sterile vacant lot. Our only realistic hope left of finding living organisms and intelligent civilizations is by looking out past our solar system to other planets circling other stars.

Stars and their planets: Eighty-five percent of the stars in the Milky Way galaxy are not single stars, like the Sun. They're gravitationally locked with other stars in multiple star systems which are made up of mostly two or three stars. Alpha Centauri, the closest star system to Earth is composed of three stars orbiting each other. The closest individual star to Earth, Proxima

Centauri, one of the three stars in Alpha Centauri, is 4.6 light-years or 26,000,000,000,000 (26 trillion) miles away. It's highly unlikely that planets circling any double star, or the stars of a three-star system, could harbor life, as they couldn't possibly remain in a steady life-friendly orbit. Further, there isn't such a thing as "just" when discussing the distance between the Earth and other stars, as in "just 4.6 light-years away". Even the closest stars to Earth are so far away they almost might as well not exist for us.

Distance and size proportions when discussing planets and stars are fascinating. If the Sun were the size of a tennis ball, the Earth would be the size of one half of a grain of rice, 9 feet away. Proxima Centauri, would be the size of a golf ball, over 5,000 miles away, greater than the distance from Los Angeles to Washington DC. Just imagine the kind of telescope that would be required to see a golf ball from that distance.

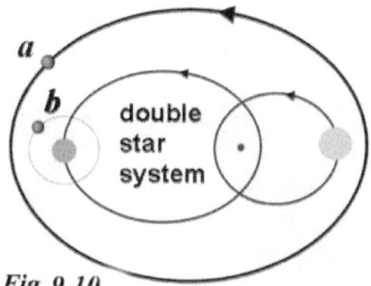

Fig. 9-10

You can see from Figure 9-10 that planets a and b, orbiting a multi-star system, would have greatly varying temperatures. They would go from raging hot when they're close to one of its stars, or when both stars are too close, to unimaginably cold. The two or three stars would make the temperature on any planet orbiting them way too variable to have any hope of being able to support life. On rare occasions, planets orbiting two-star systems might possibly support life. If two stars orbit each other with large separations, they formed independently and are called a *wide pair*. If the two stars are very close, and transfer matter from one to the other by tidal forces, then they are called a *close pair*. It might be possible for planets orbiting one star in a wide pair system to support life if the second star is so far away that it doesn't cause the planet to get so hot that any budding life is sterilized. If the stars were exceedingly close like they are in Figure 9-11, they would act like a single star, and not have sterilizing effects on planets circling both of them. Actually, it's highly unlikely that life might initiate or survive on any

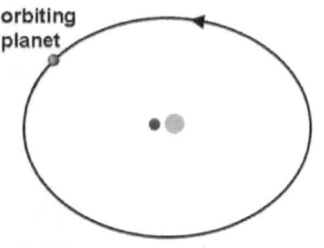

Fig. 9-11

planet circling any multiple star systems, which eliminates 85% of stars as candidates that would support life.

There are three major ways astronomers have of discovering *exoplanets* (planets outside of our solar system):

(1) **By direct imaging:** This can be done only with exceedingly large exoplanets that are relatively close to Earth. No planet that can be photo-imaged is capable of harboring life. These exoplanets are usually going to be gas giants like Jupiter and Saturn.

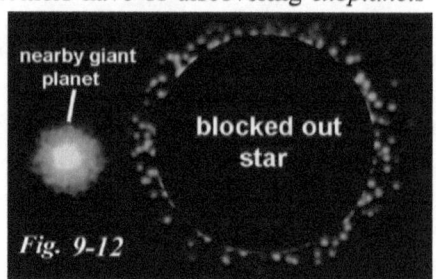

Fig. 9-12

By covering the disc of light from the star the exoplanet orbits, (Figure 9-12) the exoplanet can be directly photographed with very powerful telescopes. To date, thirty-three planets have been imaged in this way. The pictures are very pixilated, and show no details. It's not expected that there are many more planets that will be directly photographed. Few planets are close enough to Earth to allow direct imaging.

(2) **By regular dimming:** If the planet's orbit is horizontal to our view on Earth, when the planet circles between its star and the Earth, the light from the star will dim. (Figure 9-13) If this dimming occurs on a regular basis,

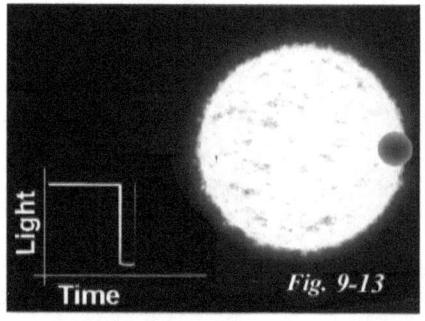

Fig. 9-13

an orbiting planet may be the source. Calculations can yield the mass of the planet and its time of orbit around the star. In May of 2016 NASA announced the Kepler mission has discovered 1,284 planets using this technique, the most exoplanets announced at one time. This more than doubled the number of previously confirmed planets. Sadly, these planets will never be directly viewed, visited, or photographed. They are way too distant.

(3) **By star wobble:** Imagine watching two children, one larger and one smaller, facing each other and holding hands. They swing around each other, each making a circle. The bigger

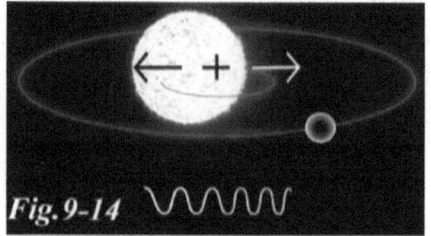

Fig.9-14

child makes a smaller circle, and the smaller child makes a larger circle. You would see each child alternately approaching and receding from you as they circle each other. This is very similar to what planets do to the suns they orbit. If the planet is massive, and its orbit is tipped and not horizontal from Earth's point of view, it causes the star to wobble enough so that wobble can be observed from Earth and our space telescopes. (Figure 9-14) This wobble can be graphed, and information can be gleaned about the planet. To date, five hundred and thirty-six planets have been discovered by this means. [19,20]

There are 53 stars within 15 light-years of Earth. Only two are sun-like stars. One of those is Proxima Centauri, a highly unlikely candidate for life because of its triple-star system. The other is Tau Ceti. Most of the others are white dwarfs, red dwarfs, and red giants. None of these can support life-friendly planets. If these stars are a good sample of all stars in the Milky Way, 95% are completely eliminated from being potentially life-friendly; which sounds discouraging. But 5% of the stars are possibly life-friendly, which represents as many as five billion stars; still an immense number. [21]

Red dwarfs make up 75% of stars in the in the Milky Way Galaxy. They are far dimmer than the sun, way too dim to be seen with the naked eye from Earth. They are about 7.5% to 50% the mass of the sun, which causes their core and surface temperature to be only several thousand degrees, far lower than the temperature of the sun. The sun, by comparison, can reach up to 27,000,000°F at its core, and 10,000,000°F on its surface; on a hot day. The low temperatures on red dwarfs also mean that they burn through their supply of hydrogen far more slowly than do sun-like stars. While other more massive stars only burn through the hydrogen at their core before coming to the end of their lifetimes, red dwarfs consume all of their hydrogen, in and out of their core. Astronomers have computed the lifetime of red dwarfs to be trillions of years, far beyond the 10-billion-year lifetime of sun-like stars. Yes folks, in about five billion years, our Sun will become a red giant star. It will expand past the orbit of the Earth, which will turn our home planet into a crispy piece of bacon. It will be the ultimate of global warming. I hope Al Gore isn't somehow here to see it. Just imagine the speeches he would be making. When the Sun finally expires, its core will collapse. It will become a white dwarf star, which is about the size of the Earth with over half the mass of the sun. A teaspoon full of a white dwarf would weigh three to four tons. [19-21]

Scientists had always thought any planets orbiting red dwarfs would be unfit for life. The low temperatures of these stars mean that their habitable zone would be very close to the star. Red dwarfs are less than 3% as bright as the sun. Consequently, any planet orbiting a red dwarf at a distance that would allow for liquid water would have to be closer than 27 million miles. (We are about 93 million miles from the sun.) A year for a planet circling a red dwarf would be less than six days. Just think. Christmas would come every week! The shopping malls on these planets would be constantly jammed. A number of planets have been discovered orbiting red dwarfs. Radiation would inundate any of these planets in the habitable zone, and fry any chance that life might attempt to arise. Further, planets so close to the star would be tidally locked, much like the moon is tidally locked to the Earth. Tidally locked means one side of the planet would be constantly facing the red dwarf. This would again be a life killer. The side facing the red dwarf would be very hot, while the side facing away would be incredibly cold. Planets circling red dwarf stars are not places where happy healthy cells might develop. Red dwarf stars are frequently covered by *starspots* that can dim their emitted light by up to 40%, for months at a time. Can you imagine the effect on the Earth if the light from the Sun were reduced by 40% on a frequent basis? Some astrobiologists think life may survive using some sort of hibernation, or possibly by diving into deep water where temperatures could be more constant. After the cold has ended, the planet's frozen oceans would reflect light like snow and ice does on Earth, greatly reducing planetary temperatures. Could a planet even hold water with such overwhelming activity from its nearby star? It's unlikely. Red dwarfs also emit gigantic flares that can double their brightness in a matter of minutes. These flares, with their immense variation in brightness and radiation, could quickly sterilize any form of life on nearby planets. Flares produce massive amounts of charged particles that could strip off sizable portions of the planet's atmosphere, if one existed. More likely, flares would prevent planets from ever having atmospheres. The fact that red dwarfs make up 75% of the stars in the Milky Way Galaxy, and the fact that they are universally so inhospitable to life, significantly decreases the possibility of astrobiologists finding signs of life on any nearby star.[22-28]

But ever-optimistic astrobiologists ignore these facts. They have now changed their minds, and say that new models show planets orbiting red dwarfs may be capable of accommodating life. The first planet to be found circling a red dwarf, Gliese 581g, was discovered in 2010. It was called the "first potentially habitable" exoplanet. Why is it potentially habitable? Gliese 581 is located "just" 20 light-years away. To give you an idea just how far Gliese

581g is, it would take a spacecraft traveling at 35,000 miles per hour 383,217 years to reach it. If Gliese 581g were proportioned down to the size of a grain of rice, it would be as far away as the moon. Its atmosphere, if there were one, would be undetectable. The amount of information we can ever glean from such a "nearby" planet that is really such an incredible distance away will forever be puny. It's also very safe to assume that humans will never explore any planets outside of our solar system. Any random radio waves coming from a planet that far away would be so weak that they would be useless or undetectable. The only chance of picking up radio waves would be if an advanced civilization, on Gliese 581g or any similar exoplanet, purposefully beamed them toward Earth. But they would need an incredibly powerful radio beam, aimed at, from their vantage point, our nearly invisible planet. The notion that there are aliens on a nearby exoplanet, that they are intelligent, that they could find us, or have any interest in us, and they would aim an incredibly strong radio beam our way is pure illusion.

If there are intelligent civilizations on exoplanets, and they aren't radio capable, we will never know anything about them. If they are very advanced, and capable of receiving and sending communicative signals, the radio capable civilizations must be within 15 or 20 light-years of Earth. Imagine for a moment that Gliese581g had an intelligent civilization that was bent on communicating with us Earthlings. If we received some kind of radio waves from Glieselings, it would mean they sent the waves 20 years ago. What might be the lifespan of the average Glieseling? If it were short, like 20 or 30 years, the Glieselings that sent the signals might be dead by the time we received their message. It would take time for astrobiologists to translate the message, let's say a couple of years, before we could send a response. We would have to build an immense radio beacon capable of reaching them with a signal, which would also take years. Once sent, it would take twenty years for our message to reach the Glieselings, and maybe a few years for them to translate our message into Glieselingese, or whatever their language might be named.

Just imagine what our communication with Glieselings might be after our SETI radio antennae received their alien message. After a few years of decoding, astrobiologist decoders might translate the Glieselings message and find something like this: "Hi! We from planet way way far away! Who you?"... A few years later after translating their language, we might send back, "Hi yourself! We are Earthlings! We call you Glieslings. Are you Glieselings peaceful people?" Then, forty-two years later, the Glieselings might retort:

"Some of us are..." ZZZZZZZ With each cycle of outgoing communication from NASA, and response from the Glieselings, half of the population of the Earth who witnessed each outgoing message would be dead; mostly from old age. Of course we could inundate them with information about who we are and what we do, and they could do the same for us. But the enormous job of decoding alien language without any points of reference may be extremely difficult if not impossible. If we received some kind of message, it's likely we may never be able to translate their code at all. Actually, the impossible part is locating radio waves from an intelligent civilization with the use of giant antennae. Just think how difficult it is to receive clear radio waves from sixty miles away on Earth. If we received signals from a civilization 100 light-years away, it would take a minimum of 200 years between each question and response cycle. No person on Earth who witnessed an attempt at communicating with the aliens would be alive for the response, if it did come. Any information they sent us would most likely be obsolete by the time it reached us. Consider the advances in technology that occurred on Earth in the last hundred years. It seems that if we Earthlings ever do find telltale signs of an intelligent alien civilization, we will never be able to get much information about who they are and what they do. Many human generations would elapse before any reasonable communication might be established. Sometimes this stuff is fun to think about and sometimes it just isn't. This subject is both fun and disappointing.[29]

Astrobiologists had a real "say hallelujah" moment when, in July of 2015, NASA announced it had discovered Earth's "bigger older cousin", a rocky planet they named Kepler 452-b. This was a real breakthrough, as until this point, there were no rocky planets discovered that were within the habitable zone of any "nearby" star. Kepler 542-b was the first. Until Kepler 452-b there were no known "nearby" stars like our Sun that were capable of supporting life. Kepler 452-b's Sun is like ours. Kepler-452-b orbits its sun-like star every 385 days, almost the exact same 365-days it takes Earth to trek around our sun. Kepler 452-b is probably a rocky planet about 50% larger than the Earth. Its mass is 5 times that of the Earth, so it would have a pretty strong gravitational pull. We would weigh at least twice what we do here on Earth. Which means any intelligent beings inhabiting Kepler 452-b would be vastly different than humans.

Figure 9-15 is a NASA artist's rendering of Kepler 452-b. NASA released a statement that said,

Fig. 9-15

This is about the closest twin to Earth 2.0 that we've found so far, and I really emphasis so far... The exciting thing is we've found so many planetary systems that are unlike our solar system. Signs of life require advanced technology and instruments in space, and so what Kepler is doing is telling us there are world's out there that we can go forward and pursue following them up.

"So far" means that this is just the beginning of huge expenditures, so wachout you taxpayers! According to NASA, Kepler 542-b's Sun is 4% larger, and 10% brighter than our sun. The atmosphere of Kepler 452-b should be far thicker than ours due to its gravity. Because of its mass it should have very active volcanoes. NASA's press release is designed to allow more funding so they can find more planets that have an infinitely small chance of harboring life, and zero chance of having that life detected by our scientists. The illusion that we can and will find life in the universe must be promoted at all costs. NASA's scientist continued,

You and I probably won't be travelling to these planets; but our children's children's children could be. This gives us something to aim for. One generation from now we might be able to get there... Before we get pictures of the sorts we just got from Pluto, where we flyby a distant exoplanet and beam back to Earth the images, the speed at which the aircraft can get there, the speed of light, is many decades (away)... If we had a sufficiently large telescope with current technology we could make the first primitive maps of an Earth-like planet. Whether they have oceans, clouds, perhaps even seasons... Whether we can discover life is a tricky question. Would we recognize those signs of life? Those kinds of telescopes might be able to do that as well.

You see, NASA must keep our hopes up so we taxpayers will be happy to continue funding their projects. Searching for planets that can harbor life, and determining if any planets have life, has no chance of success. The NASA scientist that concocted this statement must know that even at the speed of light it would take 1,500 years to get to Kepler 452-b. And if it did get there, it would take 1,500 years to for humans on Earth to receive a signal from the spacecraft announcing its arrival. By that time no one would be aware of the project, or even care. If the spacecraft left Earth today at light speed, we wouldn't find out its fate for 3,000 years; not until the year 5015. This is longer than the span of time from King Tut until now. We won't ever do a flyby discussed by the NASA scientist; unless we want to send a camera that will take 3,000 years to report back its exploits and send photos. Of course, the instrument holding the camera must attain light speed to even report back in 3,000 years. The NASA scientist who wrote this press release must know this. In fact, of course he does. Which makes him nothing but a scientific illusionist.

To their credit, the discovery of Kepler 452-b by NASA is nothing short of astounding. If Kepler 452-b were the size of a one-inch marble, it would need to be placed 11,835,000 miles out into space to be equivalent. Or it would be like looking at a tenth of a grain of rice on the moon. This should give you an idea of how absurdly far it is from Earth. If we could send a rocket ship to Kepler 452-b that went 100,000 miles per hour, the fastest any space vehicle has ever traveled, it would take over ten million years to get there. Our kids won't be making this trip, nor our kid's kids, nor any descendant we will ever have. Mankind will never go to Kepler 452-b. Never. We will never build a telescope that will be able to directly view it. For this NASA scientist to say we will someday travel there, and even view and map it with a telescope, is nothing but an illusion. Equating Kepler 542-b with Pluto is illusion at its best. Kepler 452-b is 2 million times farther away than Pluto. Our probe took 10 years to get to Pluto. At the same speed, it would take 20 million years to reach Kepler 452-b. The NASA scientist is simply too educated and smart to not know how absurd his statements really are, or he wouldn't be working with NASA. For NASA to allow the release of that statement makes NASA nothing but a scientific illusionist government organization. Kepler 452-b will never be anything more than a planet that causes a star that is very far away to either wobble, or dim slightly on a regular basis. We will never photograph Kepler 452-b. No one is more disappointed than I am. I would love it if there were a way to directly view Kepler 452-b, or even visit it. But the universe has placed a roadblock in our way. That roadblock is unimaginable distance; and that roadblock is impossible for mankind to pass. [30]

Even though the distances to other stars are so immense they make our search for life incredibly expensive and futile, scientists have been able to devise mathematical formulas that at least give us an imaginary idea if life exists on exoplanets. In the 1960's, radio astronomer Dr. Frank Drake developed an equation for the purpose of calculating the potential number of intelligent civilizations in our galaxy, the Milky Way. There is no sense ever considering any other galaxies in our search for life, as they're so unbelievably distant. Humans will never know anything about the existence of any solar system in any other galaxy but our own. The nearest major galaxy to ours is Andromeda, at 2.2 million light-years away. This means light, travelling from Andromeda at 186,000 miles per second, takes 2.2 million years to reach Earth. So if you look at an image of Andromeda, you are looking 2.2 million years into the past.

Since we will never detect a civilization close enough to Earth to have any communication with, Drake's formula is really just an imagination amplifier. But it's certainly a fun one to think about. We've learned a great deal since Dr. Drake made his equation. In the 1960's, astronomers had no idea if there really were planets circling other stars. Since, using advanced techniques, telescopes, and satellites, astronomers have discovered hundreds of exoplanets. The discovery of exoplanets makes Dr. Drake's formula far more interesting. It gives a mathematical estimate of the number of active, intelligent extraterrestrial civilizations that exist in the Milky Way Galaxy. His formula goes like this:

$$N = R^* \times f_p \times n_e \times f_l \times f_i \times f_c \times L$$

Where:

N is the number of *radio wave capable* intelligent civilizations in our galaxy

R^* is the number of stars in the galaxy

f_p is the fraction of those stars that have planets

n_e is the average number of planets that can potentially support life per star that has planets

f_l is the fraction of the above that actually go on to develop life

f_i is the fraction of the above that actually go on to develop intelligent life

f_c is the fraction of civilizations that develop a technology that releases detectable signs of their existence into space

L is the fraction of the lifetime of a planet that its technical civilizations release detectable signals into space.

If you don't remember the math, here is a review: The odds of rolling a 6 with a dice three times in a row is found by multiplying the odds of getting a 6 on each roll times itself 3 times. In other words, $1/6 \times 1/6 \times 1/6 = 1/216$; or one chance in 216 tries. Drake's formula works by multiplying the odds of his scientific variables in the same way. People have plugged in a variety of values to Drake's equation over the past 50 years — all of them purely speculative of course. Values for N have ranged anywhere from one civilization into the millions. The late Carl Sagan, in his renowned series, *Cosmos*, gave his estimated solution to the Drake equation. No astronomer was more respected than Carl Sagan, so I thought it would be interesting to use his numbers to come up with a good estimate. The truly amazing thing about Sagan in this documentary is he treats the Drake equation as if it's absolute truth. He spends no time determining or discussing if the equation is valid. He treats it like it's on par with $E=mc^2$ or $A^2 + B^2 = C^2$. Wouldn't a good scientist *validate* the equation first? It's as if the solution Sagan will get is completely valid, without question. He keeps a running total as he goes. Things he has to say about his estimates are in italics. In parenthesis are my comments. Here are Sagan's estimated numbers:

$R*$=400 billion stars in the Milky Way Galaxy (Me: A very high estimate. 100 billion might be more realistic. 90% are eliminated because they are red dwarfs, and/or double and triple stars. White dwarf stars, red giants, and other non-life-friendly stars further reduce Sagan's number. Sagan must have known this, so why didn't he take it into consideration? His number should be 100 billion reduced by 90% = 10 billion. But I will follow Sagan's plug-ins as if they are correct.)

f_p = 1/4 of the 400 billion stars in the Milky Way have planets=100 billion stars with planets

n_e = 2 planets per star system that can potentially support life= 200 billion planets that can potentially support life. Sagan: *There should be 100 billion planets times ten, or a vast arena for the cosmic drama. In our own solar system, there might be several bodies suitable for life, life of some sort...There is the Earth, of course, but there are possibilities for Mars, for Titan, for Jupiter...* (Me: Sagan "conservatively" chose 2 worlds per star that are suitable for life. Astrobiologists now know that there are no other sites in the solar system besides Earth capable of supporting life, so his conservative estimate should be 1 planet per star not 2.)

f_l = ½ of the 200 billion planets that can potentially support life = 100 billion inhabited worlds that actually go on to develop life. Sagan: *Now what about life? Under very general cosmic conditions, molecules of life are readily made, they spontaneously self assemble. It's conceivable that there may be some impediment, like some difficulty in the genetic code, although I think that's very unlikely, given billions of years for evolution. On the Earth, life arose very fast after the planet was formed.* (Me: Sagan chose one half for the planets capable of supporting life that actually developed life? A very generous estimate. According to biologists, life arose 800 million years after the formation of Earth, not so "very fast". Molecules of life are NOT "readily made", nor do they "spontaneously self assemble". It's hard to believe such an intelligent educated person like Sagan didn't realize these facts. Or did he?)

f_i = 1/10 of the 100 billion planets that form life develop intelligent life=10 billion planets that develop intelligent life.

f_c = 1/10 of the 10 billion planets that develop intelligent life develop a technology that releases detectable signs of their existence into space= 1billion planets that release detectable signs of their existence. Sagan: *planets on which civilization arose at least once.*

f_L = 1/100,000,000 = the fraction of the lifetime of a planet that is marked by a technical civilization = 10 planets that harbor intelligent civilizations that release detectable signals into space during a planet's existence.

Sagan: *What percentage of the lifetime of a planet is marked by a technical civilization? The Earth has harbored a civilization capable of radio astronomy only for a few decades out of a few billion years. It's hardly out of the question that we might destroy ourselves tomorrow. If that were a typical case, then f_L would be a few decades divided by a few billion years, or 1/100,000,000. (one hundred-millionth). Then N would be one billion divided by one hundred million, or... N may be just ten. A tiny smattering, a pitiful few technological civilizations in the galaxy. There may be a few others, there maybe nobody else to talk to.*

(Me: A pretty honest conclusion. If he knew how difficult it is to get any spontaneous assemblage of biochemicals, his N would equal 0. But 10 is a reasonable solution. Ten out of four hundred billion stars might as well be zero. But hold the phone. Sagan really blew it. He gave up his semi-conservative math for imagination and conjecture.) He goes on: *But consider the alternative. That occasionally civilizations learn to live with high technology and survive for geological or stellar evolutionary time scales. If only one percent of civilizations can learn to survive technological adolescence, then f_L would be not one hundred millionth, but only one hundredth.* (Me: How on Earth did he come up with this number?) *And then the number of civilizations would be a* *billion times 1/100. The number of civilizations in the galaxy would be measured in the millions... millions of technical civilizations.*[31]

Did Sagan make the leap from ten and possibly zero to millions of technical alien civilizations to keep his audience excited about the possibilities? Of course he did. Everyone that has even a tiny bit of interest in this subject wants there to be aliens all over the universe. It's such an exciting thought. Sagan's imagination destroyed his fairly honest math. But, to his credit, he made people like me happy, and relieved, by concluding that there could be millions of intelligent alien civilizations. That's what everybody wants to hear. Three cheers! Actually, Sagan did create an incredible amount of interest in astronomy. He really peaked my interest. So Sagan gets tons of credit for making astronomy interesting and fun. His analysis of Drake's equation was far from perfect, but no one is perfect. [31-33]

Dr. Drake's equation deals with estimating the number of intelligent civilizations in the Milky Way Galaxy that give off detectable radio signals. A professor of planetary science and physics at the Massachusetts Institute of Technology, Dr. Sara Seager, came up with her own take on the equation. Her new formula seeks to determine the chance of finding life *of any kind* on planets that are observable from Earth. She considers hers to be a parallel equation, not one that is supposed to supplant or compete with Drake's equation. Her equation deals only with the percentage of stars having planets that can and do support life. Her equation does not deal with aliens that send out detectable radio signals, as Drake's does. Her formula is:

$$N = N_* \times F_Q \times F_{HZ} \times F_O \times F_L \times F_S$$

The following are what each letter represents. Dr. Seager's estimates for their values are in parenthesis:

N = the number of planets with detectable signs of life=

N_* = the number of stars close enough to Earth to detect planets =30,000

F_Q = the fraction of stars that are quiet with low amounts of radiation and flares (0.2) = 6,000

F_{HZ} = the fraction of stars with rocky planets in the habitable zone (0.15) = 900

F_O = the fraction of those planets that can be observed (0.001)=9

F_L = the fraction that have life (1)=9

F_S = the fraction on which life produces a detectable signature gas (0.5) = .45

F_S **Fraction with Detectable Spectrscopic Signature**

$$N = N_*F_QF_{HZ}F_OF_LF_S$$

Term	M Stars
N_*	30,000
F_Q	(0.2)
F_{HZ}	0.15
F_O	0.001
F_L	1
F_S	0.5
N	2

Strangely, Dr. Seager's own estimates and calculations that are on many sites on the Internet (left) give a result of 2 stars with planets that give off detectable signs of life. Dr. Seager states that the first four numbers are pretty well locked in. The last two are just estimates. She says her calculations suggest 2 inhabited planets might actually be located during the next decade. I think she hit the wrong button on her calculator. Using her figures and estimates I get a result of 0.45, not 2. But 2 out of 30,000 possibilities is really the same. Being the skeptic that I am, I would say that finding hints of life on two planets out of 30,000 is the same as saying there is no findable life on 30,000 planets. If 2 planets out of 30,000 have life, they will never be found. Zero life supporting planets is just as good of a number as 2, and far more realistic. It really doesn't matter if Dr. Seager's result is a mathematically incorrect 2 or a correct 0.45, as you shall see.[32-34]

Of course I have my own thoughts on Drs. Drake and Seager's equations. Both are in the wishful thinking category. But that's good in a way. Everyone who thinks about this stuff wants there to be alien civilizations. I certainly do. What a boring and self-absorbed thought to think the universe is made for only us Earthlings; an absurd notion. Both equations give a completely distorted and hugely overly optimistic notion about whether or not there is extra-terrestrial life. Both Dr. Drake's and Dr. Seager's equations are missing some immensely important variables. Are they written to keep excitement up for their projects; and funds coming in as well? Or simply to keep people excited? Whatever the reason, I rewrote both equations, adding a bit of reality; actually a lot of reality. I do have to give Dr. Drake credit for thinking up the idea of putting together a formula for figuring out the chance of intelligent life in the galaxy. That's an excellent idea. He just didn't think it through as much as he should have. And credit goes to Dr. Seager for her equation. It will never be known if Dr. Drake's equation is valid. It's highly unlikely that it will ever be known if Dr. Seager's is.

Here is the problem with both equations. There are hundreds of variables that neither account for. If I plug in the most basic of those variables that both Drs. Drake and Seager missed, both equations collapse. What Drs. Seager, Drake, and Sagan missed is the fact that if any of the *individual* variables of either equation approach zero, their formulas are zeroed out. I am going to show that variables that must be embedded in both equations do just that. Which means that the chance of life forming on any other planet, not only in our solar system, but in the entire universe, is zero. As I said, there are many variables that could wipe out both equations; but I am only going to use a few.

Any life-friendly planet circling a star must have a powerful magnetic field, a *magnetosphere*, surrounding it. A magnetosphere protects a planet from the immense radiation storms and star winds that would kill off life that may be in the process of forming. The Earth is the only known planet with an immense magnetic iron core. It's not quite like a bar magnet, however. The Earth's magnetic field changes over time because it's generated by molten iron, which is in a constant state of flux caused by the motions of the Earth. Our magnetosphere is immense, and it protects us from killer solar radiation and winds. If we conservatively estimate that one in four rocky planets have a significant magnetosphere, as is the case with the solar system, at least 75% of all rocky planets can be eliminated as possible sources for life. The truth of the matter is that far fewer planets have significant magnetospheres than one in four. But one in four will do as a very conservative estimate. Potential life-supporting planets are now reduced by 75%. I will label the need of a magnetosphere f_m.

A planet that may be a candidate for the formation of living organisms must also have a healthy atmosphere that is amenable to promoting life. A healthy planetary atmosphere will further protect a planet and its living organisms from radiation. The Earth's atmosphere is perfect for life. The thick atmosphere on Venus would kill off any living organisms. The atmospheric pressure on Venus is 90 times greater than the Earth's. It has a cloud layer composed of sulfuric acid, which constantly rains down acid on Venus' surface. Its atmosphere acts like a greenhouse, which makes Venus' surface temperature hot enough to melt lead. The atmosphere on Venus makes it an incredibly efficient sterilizer, far better than any medical sterilizer on Earth. Mercury has almost no atmosphere, and it's the closest planet to the sun. Because of this fact, Mercury's surface has the greatest temperature variation of all planets, ranging from −280°F at night to 800°F during the day. Interestingly, Mercury is 36 million miles from the sun. Venus is almost twice as far at 67 million miles. But Venus is hotter due to its greenhouse effect. The average temperature on Venus is 864°F. The atmosphere on Mars is only 0.6% as dense as the Earth's. Without a magnetosphere, and with such a thin atmosphere, Mars surface is inundated with killer solar radiation. Again, using the solar system as a model, only 25% of rocky planets have a life-friendly atmosphere. I will label a life-friendly atmosphere as f_a. So f_m x f_a =.25 x .25 = .0625 or, very conservatively, 6.25% of rocky planets have both a magnetosphere and a life-friendly atmosphere.

Other major variables that Seager, Drake, and Sagan missed in their equations have to do with biology and biochemistry. On Earth, all living organisms are composed of cells; without exception. Earth is an incredibly life-friendly planet. It's unimaginable that any of Dr. Seager's 30,000 stars will have a planet that will be a better location for life than the Earth. Eighty percent of the Earth is covered with water; it has plentiful carbon, and all the other elements and molecules that life requires. Silicon has many of the properties of carbon. If there were any other kinds of life possible, say life based on some other element, like silicon instead of carbon, it would have formed here. It would be easy for scientists to observe. Why, then, are there no silicon-based life forms on Earth, especially given that, on Earth, silicon is 135 times more abundant than carbon? If life could be silicon-based, it sure would show up here. But it doesn't. So it must be assumed that life on other planets would have to be based on water, and carbon, like it is on Earth. A variable must be included that addresses the planet having an ample and usable supply of carbon and water. I will label the water/carbon variable f_{wc}. This variable asks, "What fraction of exoplanets have a plentiful presence of both carbon, and liquid water?" The only planet with abundant water in liquid form in the solar system is Earth, out of the four rocky planets. Carbon is fairly abundant on other planets. Carbon dioxide is present in the atmosphere of Venus. Methane (CH_4) is a common ingredient of the atmosphere of many planets. Europa, Jupiter's moon, has plentiful water ice, and possibly some liquid water. It's unlikely one in fifty planets will have both available carbon, and plentiful water in liquid form. But I will generously give f_{wc} a value of .25. So my addition to Drs. Drake and Seager's equations becomes f_m x f_a x f_{wc} = .0155625. Or, about 1.5% of rocky planets have a magnetosphere, life-friendly atmosphere, and plentiful carbon and liquid water. [35]

Cell membranes, the enclosures of living cells on Earth, are made up of about 50% lipid (fat) molecules, and 40% proteins. Both lipids and proteins are products of living organisms, which begs the question, what was the source of lipid molecules and protein molecules that formed the first cell membranes on Earth? Ignoring proteins for a moment, there needs to be a supply of lipid molecules for cellular life to begin. Scientists say the first cells on Earth came from tiny bubbles, or micelles, made of lipids. Going from a sterile planet, flush with water and carbon, to one with living cells requires lipids. Do lipids exist on any of Drs. Seager and Drakes exoplanets? Answer: no. Lipids are manufactured by living organisms. They don't exist on any sterile planets. So both equations need the addition of a variable that determines how many exoplanets have lipid molecules. The answer here should be zero, which completely kills both equations. For the good of astrobiology, and to keep the discussion going, I'll add an f_{lip} to both equations, and generously give the presence of lipids a .10% chance of existing on any planet before the formation of living cells. I doubt any scientist would argue this generous value. Our newly calculated estimate of the percentage of planets that have the capability of forming life now reads:

$(f_m \times f_a \times f_{wc} \times f_{lip})$ =.25 x .25 x .25 x .001= 0.000015625 or 0.001562%; a little more than a thousandth of a percent. Plugging these four variables into Sagan's estimates of Drake's equation yields:

f_p = 1/4 of the stars in the Milky Way have planets=100 billion stars with planets

Adding in my very obvious, realistic, and necessary variables:

$n_e = (f_m \times f_a \times f_{wc} \times f_{lip})$ = 0.001562% of the 100 billion planets can potentially support life = 1,562,000 planets

f_l = ½ of 1,562,000 inhabited planets that actually go on to develop life = 781,000 planets.

f_i = 1/10 of 781,000 planets develop intelligent life = 78,100 planets that develop intelligent life.

f_c = 1/10 of 78,100 planets that develop intelligent life develop a technology that releases detectable signs of their existence into space = 7,810 planets that release detectable signs of their existence.

f_L = 1/100,000,000 of the lifetime of a planet is marked by a technical civilization = 1/100,000,000 of 78,100 = 0.00781 or less than one-hundredth of one planet in the Milky Way Galaxy harbors intelligent civilizations that release detectable signals into space during a planet's existence. In other words, zero planets.

Dr. Seager's equation calculates like this with the addition of my variables:

N = the number of planets with detectable signs of life =

N_* = number of stars close enough to Earth to detect planets (30,000)=30,000

F_Q = fraction of quiet stars with low radiation and flares (0.2) = 6,000

F_{HZ} = fraction of stars with rocky planets in the habitable zone (0.15) = 900

(F_M x F_A x F_{WC} x F_{LIP}) = 0. 001562% of the 900 rocky planets have a magnetosphere, life-friendly atmosphere, usable water and carbon, and free-floating lipids = .01458 planets

You can see that the last three steps in Dr. Seager's equation can be eliminated. We are already essentially at zero planets before we reach planets that can be observed (F_O), have life (F_L), and produce detectable signature gas (F_S).

As you can see, adding only a few realistic variables into Drake's equation with Sagan's solution, and Seager's equation causes both to completely collapse. We won't detect nearby planets that show signs of life, nor are there planets anywhere in the Milky Way with intelligent life.

A final killer of both the Seager and the Drake equations, certainly not the last variable, is the chance formation of proteins. Proteins are the building blocks of all living cells. Living cells are the building blocks of living multi-cellular organisms. Without proteins, there can be no cells, and no life. Again, on the planet Earth, there are no living organisms whose cells are not based on proteins. About a quarter of the weight of all living cells are proteins. If life were possible without proteins, it would be present on life-friendly Earth. Just as there is no life on Earth not based on carbon, there is no life not based on proteins. Humans have at least 90,000 different very specific proteins in their bodies. To simplify things for astrobiology, I'll cite just one average-sized protein of those 90,000. If a planet that has water and carbon, and can form some kind of lipid cell capsule, and by a miracle of dumb luck form *one single* protein molecule, in this imaginary tale, life may be off and running. (Actually not, as we don't have any idea what life is, or what it's source is. Life cannot be synthesized. No entity with water, proteins, and lipids, comes to life. But in this illusion, let's pretend that with these few ingredients it does.) The only problem here is the building blocks for proteins are amino acids. On Earth, over 300 amino acids exist in nature. Only twenty-one of these are utilized in living organisms. For life to have formed on any exoplanet, those twenty-one amino acids had to isolate themselves somewhere, by **D**umb **L**uck happenstance, so they could assemble themselves in a very specific order to form the first usable protein molecule of many billions that are needed by living organisms. Ideally this isolation took place inside of a lipid cell capsule. Living cells manufacture amino acids, but that fact must be ignored in this discussion, or there is no discussion.

I explained in great detail in *Evo-illusion* about the odds of a protein molecule self-assembling. Any average protein could be used as an example. I used *hemoglobin*, the protein that carries oxygen to our cells, as an average-sized protein molecule. Hemoglobin is made of four strands of amino acids. Taking only one strand of the four, and calculating the odds of amino acids self-assembling into that *one single strand* yields a number that is so immense that famed physicist and evolution supporter Isaac Asimov named it the *hemoglobin number*. The hemoglobin number is $1:10^{190}$. Yes, that's one in a one with one hundred ninety zeros after it. One trillion has twelve zeros. All of the atoms in the universe total about 10^{80}, a one with eighty zeroes. Essentially, the chance formation of *one single strand* of a single four-stranded protein molecule completely and randomly self-assembling alone negates both Dr. Seager and Dr. Drake's equations. There is no need for my four variables to zero out both. Remember, $1:10^{190}$ is the odds of the chance formation of just *one strand of one average-sized four-stranded protein molecule*. If all of the four strands of an average single four-stranded molecule were calculated, the number becomes so absurd, it's not worth discussion; which is the case anyway. If you're curious, it's $1:10^{760}$. If all of the 90,000+ proteins in the human body are taken into account, the odds against random protein self-assembly approach one in infinity; which it does with only the single strand of the one average four-stranded molecule.

I could go on and discuss other variables missed by Seager and Drake. But that would be a worthless endeavor, because only the four I cite collapse both equations. A single protein molecule alone does as well. Adding more variables will just make them even more zeroed out, which is a waste of time.

I emailed Dr. Seager and asked her why she didn't include the need for some kind of lipid cell capsulation, the need for water and carbon, and the need for proteins in her equation: and why she got 2 instead of .45. Her response was:

Hi Steve

It's good to hear of your interest. The equations (as you know) are just illustrative and anyone can make up their own. Mine had a specific purpose in showcasing what we plan to do with the MIT-led TESS Space Mission to find a pool of rocky exoplanets that the JWST will follow up on.

We will have the capability of finding signs of biosignature gases, if we are really, really lucky. In that case we don't worry about the biology part, just assuming things worked out how can we detect it. Good luck with everything.

Best regards

Sara Seager, Massachusetts Institute of Technology

Professor of Planetary Science

Professor of Physics

The *Transiting Exoplanet Survey Satellite (TESS)* (Figure 9-16), which Dr. Seager writes about, is scheduled to launch in 2017. It will be rocketed into an orbit around the sun, one million miles closer to the stars than is the Earth. According to the Massachusetts Institute of Technology TESS website:

Fig. 9-16

The Transiting Exoplanet Survey Satellite (TESS) is an Explorer-class planet finder. In the first-ever spaceborne all-sky transit survey, TESS will identify planets ranging from Earth-sized to gas giants, orbiting a wide range of stellar types and orbital distances. The principal goal of the TESS mission is to detect small planets with bright host stars in the solar neighborhood, so that detailed characterizations of the planets and their atmospheres can be performed. TESS will monitor the brightnesses of more than 500,000 stars during a two-year mission, searching for temporary drops in brightness caused by planetary transits. Transits occur when a planet's orbit carries it directly in front of its parent star as viewed from Earth. TESS is expected to catalog more than 3000 transiting exoplanet candidates, including a sample of 500 Earth-sized and 'Super-Earth' planets, with radii less than twice that of the Earth. TESS will detect small rock-and-ice planets orbiting a diverse range of stellar types and covering a wide span of orbital periods, including rocky worlds in the habitable zones of their host stars.[36]

The *James Webb Space Telescope*, what Dr. Seager calls called JWST, will be a large infrared telescope that will also be sent into solar orbit about one million miles farther from the Sun than is the Earth. The JWST should be launched in 2018. It will be the premier observatory of the next decade, serving thousands of astronomers worldwide. It will study every phase in the history of our Universe, ranging from the first luminous glows after the Big Bang to, hopefully for astronomers like Dr. Seager, finding exoplanets capable of supporting life. The JWST will have seven times more collecting area than does the Hubble Space Telescope. Its camera will have a fifteen times greater field of view than the camera on the Hubble. The JWST will bring in some astounding images. The Hubble was successful beyond anyone's wildest imaginations. The JWST should be even more incredible.[36,37]

In light of the fact that Dr. Seager said, "The equations (as you know) are just illustrative and anyone can make up their own", I now have her tacit approval to "make up my own". So I will. I guarantee mine are far more accurate than either Dr. Drake's, or Dr. Seager's. I propose the following formulas as my own, and, of course, I will, ahem, name them after myself:

Dr. Blume's Equation Regarding Alien Intelligent Life in the Milky Way Galaxy/Universe:

$$N = R^* \times f_p \times n_e \times (F_M \times F_A \times F_{WC} \times f_{LIP}) \times f_l \times f_i \times f_c \times L = 0$$

Dr. Blume's Equation Regarding Alien Life on a Planet Detectable by Telescope:

$$N = N_* \times F_O \times F_{HZ} \times (F_M \times F_A \times F_{WC} \times f_{LIP}) \times F_Q \times F_L \times F_S = 0$$

By adding in my four new variables, I can now claim both of these equations to be mine. They are significantly different than the originals, and the solutions are as well. If anyone charges me with stealing someone else's work, I could simply add more valid variables; as many as I need to make them more mine. More variables will only result in the exact same solutions: zero for both equations. By plugging in the chance formation of one single strand of one average protein molecule, both of these equations are zeroed out. Mathematicians say that once you hit the number 10^{40} (one with forty zeros after it), the math is so insignificant that it becomes imaginary. The hemoglobin number, $1:10^{190}$, is light-years past imaginary.

One could argue that even though the odds approach one in infinity against life forming on another planet, it did happen here on Earth. So it *could* happen on an exoplanet. Any mathematician will give you an idea of the odds of an event with a chance of one in near infinity happening twice, once here on Earth, and a second time on an exoplanet. You multiply one over near infinity by one over near infinity, (1/near infinity X 1/near infinity) and you get… one in hyper-infinity? Scientists are looking for life in a miniscule area of the Milky Way that has a comparatively few stars that have the odds of life occurring on their nearby planets, like it did here on Earth, of one in near infinity (1:near infinity). This is a scientific absurdity. There is absolutely zero chance of there being life on any star's planetary system within telescopic sight of Earth. In fact, there is zero chance of there being life anywhere in the universe except here on Earth. The odds of $1:10^{190}$ guarantee that fact. Again, I am fudging, because a single protein molecule of any type would be completely useless. There would need to be another molecule, and another; billions upon billions of molecules of each of tens of thousands of proteins needed by any living cells. Actually when I insert the odds of a single strand of a single protein molecule forming, I am still fudging hugely in favor of the astrobiologists. A real example of fudging is ignoring the fact that *no lipid capsule with a single protein molecule inside will come to life*. The chance of that occurring is zero, adding another zeroing out of the Drake and Seager equations. If a number of events needed for the start of the building of living organisms that have an infinitely low chance of occurring somehow did occur because they have billions of years, it doesn't really matter. An unlikely physical and biochemical attempt to concoct cells through dumb luck in order to make a start at life will never come to life. Whatever happened on Earth that brought about life isn't something that could happen casually and often. The only way it happens casually and often is in the imaginations of scientists like Drs. Drake, Seager, Sagan and all astrobiologists.[38]

I shouldn't have this opportunity to name my two equations after myself. But this is *my* book, so I shall. Actually, the number of equations that would zero out both Drake's and Seager's equations is endless. There are probably hundreds of far more valid versions of these equations than theirs written by good scientists. But both equations remain at the top of the scientific totem pole. They are promoted as valid, while truly valid formulas written by other scientists are completely obscure. Reality is the notion of life on other planets, particularly intelligent life, is just a pipe dream. All we can do is guess and imagine. For us humans, life on other planets can only exist in our movies, dreams, television shows, books, and imaginations, but nowhere else.

For the sake of continued discussion, let's say life did occur on several nearby exoplanets. Let's say Dr. Seager's dream did come true. What if we could not only detect gasses that were indicators of living organisms, but what if we could collect samples of alien organisms, and study and observe them under a microscope? If that did occur, there are two possible scenarios, each with great ramifications for evolution and intelligent design, or what I call **IID** (**I**ngenious **I**nvention and **D**esign):

(1) Alien organisms are identical in most ways to living organisms on Earth. Alien life is organized much like life on Earth. There are both alien plant and animal species. The *Krebs cycle*, the very complex biochemical cycle that provides energy to all animal cells on Earth, also provides energy to alien animal cells. Alien plant cells utilize *photosynthesis*, the very complex biochemical cycle that provides solar energy to plant life on Earth. Alien cells are loaded with proteins. They have ribosomes, mRNA, and DNA that holds the plans for and manufactures those proteins. In almost all respects, alien cells are the same as cells on Earth.

The ramifications for IID and evolution: Evo-illusionists are so excited about the idea that there must be life throughout the universe. Evolution is "real science", and its proponents think this outcome would do nothing but prove evolution. They wish for this outcome; but I say be careful what you wish for. Nothing would be worse for evolution than to find alien cells that are nearly identical to Earthly cells. Cells, if they came about by evolutionary processes, would have had to go through millions of random steps that were selected for by natural selection. Because there is no possible contact or communication between any exoplanet and Earth or its cells, an alien planet that came up with Earthly cellular designs would prove the theory that there is a universal cosmic blueprint; a common design source in the universe. **IID (ID)** would be proved. Evolution would crash; badly. The random steps touted by evolution's illusionists that brought about life on Earth could not possibly occur in the exact same fashion and order on an alien planet.

(2) Alien organisms are completely different in most ways to living organisms on Earth. The Krebs cycle, and photosynthesis do not exist in their Earthly form. There are entirely different biochemical cycles and reactions that provide energy to alien cells. Alien cells are devoid of proteins, or they have similar molecules that are not made up of amino acids. They have no ribosomes that manufacture proteins, and no DNA and mRNA that hold plans for those proteins. They have an entirely different kind of coding, if they have coding at all. They use an entirely different inventory of biochemicals. Cells might be much larger than those on Earth; or much smaller. In almost all respects, the cells are vastly different than those on Earth.

The ramifications for IID and evolution: This is the outcome that evo-illusionists *should* wish for. Nothing would be better for evolution and evo-illusionists than to find alien cells that are completely different than Earthly cells. Cells, if they came about by evolutionary processes on an alien planet, would go through millions of random unguided steps that were selected for by natural selection. With absolutely no contact with Earth or its cells, an alien planet should come up with completely different kinds of cells. Vastly different alien cellular designs would be great for evolution, and disprove the theory that there is a cosmic blueprint; a common design source throughout the universe. **IID** (ID) would be the big loser; evolution would gain. Evolution touts biodiversity as its greatest proof. Biodiversity should show up in spades in alien life if evolution is valid. Evolution would still have the problem of showing how any biological entity evolved in stepwise fashion. But this outcome would be a major positive step for evolution.[38]

If we did receive communications from Glieselings, or any other exoplanetary inhabitants, it may be the most remarkable day in world history. If people from Gliese 581g sent focused radio signals to Earth, and our SETI radio dishes captured them, pandemonium would ensue. Newspaper headlines would be monstrous. Would there be ticker-tape parades for the discoverers? Fame for the astrobiologists that located the radio signals would be enormous. If simple life forms were discovered on an exoplanet due to the discovery of telltale atmospheric gasses, the reaction wouldn't be as great as the finding of intelligent beings, but it would be sensational just the same. If Dr. Seager were to be the discoverer, and I certainly hope she is, my email communication with her would be priceless. But the excitement would be fleeting, and it would die out quickly. Garnering additional information about the aliens would be next to impossible. If Dr. Seager or a like-minded scientist found signs of living organisms of some kind on an exoplanet, that's all that would ever be found. There would no doubt be a bump of excitement, and that would be it. Again, imagine anyone trying to observe a tiny chip of rice on the moon with an earthbound telescope. How much information could ever be extracted from that chip of rice from such a distance? There is a limit to what humans can do when it comes to searching for life outside of the solar system, and we have approached that limit.

As I said earlier in this chapter, I have always been so excited about the thought of life on other planets. I really started writing this chapter thinking I would pretty much agree with most astrobiologists and astronomers that the universe is teeming with life and intelligence. The more I researched, the more I wrote, and the more I thought and did mind experiments, the more I came to realize that maybe we are alone. Maybe the universe truly is for just us, a tiny speck in an infinite void; that Carl Sagan's first calculations showing ten planets in the entire Milky Way with intelligent aliens out of 400,000,000,000 stars were correct. How can anyone argue against the conclusion that there is no other life in the universe but our own?

Looking at life on Earth, in every location, in every nook and cranny, living cells are the same. All animals use the same energy cycle, the Krebs cycle, all over the Earth. Plants on Earth use *photosynthesis*. All animal cells have the same basic internal anatomy, and they have cell membranes (walls) made up of lipids and proteins. All plant cells have the same basic internal anatomy, and they have cell walls made up of cellulose. The design of all plant cells, and all animal cells on the Earth is universal, just as if a common intelligent source put them all together. If cells evolved randomly, there should be vast differences in the anatomy and biochemistry of cells all over the Earth, as they would have evolved in vastly different environments, with vastly different stepwise pathways. For a single cell, one inch might as well be a hundred miles. Cells are so tiny, that long distance travel for them is unthinkable. The first living cells couldn't go running around the Earth, going through cellular division, and leaving off their daughter cells in every location that they exist in today. The notion that ocean currents or wind spread cells over all bodies of water, and on every land surface doesn't fly. Dig down 100 feet in Antarctica, living cells are there. Enter a cave filled with deadly acid, and living cells are there. Living cells exist near boiling hot vents in the ocean, and in the digestive tracts of every animal. And every single major cell type, without exception, has the same biochemistry. If there were ever a reason to believe the universe is bubbling with life it is this: cells can't travel more than an inch in their lifetimes, but their even distribution and designs are universal on Earth. They must also then be universal in the universe. If alien cells could be found and observed, they must have the same design as cells on Earth. They must, because the design source must be and will be the same. It has to be. Pure mathematics says so. Those that are evolution supporters should wish that there were no life in the universe other than here on Earth. Because the only way extraterrestrial life can exist is if **IID** is the source. A cosmic blueprint could wipe out odds of $1:10^{190}$, or $1:10^{760}$, or.... The cosmic blueprint brushed off those odds here on Earth, and most likely did so in trillions of locations in the universe. The only way life could exist throughout the universe is if there truly is a universal design source. If abiogenesis and evolution are the sources of living cells, there isn't a chance in hell that life exists anywhere but here on

Earth. Evolution has to live with the hemoglobin number, $1:10^{190}$. A universal design source doesn't.[39]

If there is life in other locations in the universe besides here on Earth, its ramifications are *just the opposite* of what is touted by modern science. The greatest evidence for **IID,** a cosmic blueprint, is the fact that we live in an immense universe that could not possibly have been formed just for the people of Earth. What could be more absurd than the notion that the entire universe is here just for our existence, enjoyment, and amazement? The existence of that immense universe, and the absurdity of the notion that it is here just for us is immense proof that **IID** is the source of all of living nature; in fact all of nature itself. So I'm going to finish this chapter with my third law, also named after (ahem) myself, with *Blume's Law Regarding Evolution, IID, and the Existence of Extraterrestrial Life*, which states:

If abiogenesis and evolution are the sources of all living organisms on Earth, then Earth houses the only living entities in the universe. Conversely, if Ingenious Invention and Design (IID), a cosmic blueprint, is the source of all living organisms on Earth, then the universe is teeming with living entities.

Please hold your applause until I sell at least a billion books.

Chapter 10

What Will Become of Us?

There are two ways to be fooled. One is to believe what isn't true; the other is to refuse to accept what is true.- Soren Kierkegaard

Have you ever wondered what the fate of humanity on Earth will be? How long can we live on this planet? Are we humans evolving to be better and better as the years, and millennia go by? Will our descendants all be intelligent beyond our wildest imaginations due to evolutionary improvements? Will they be more skilled at every field of endeavor? If you go by what the evolution and biological sciences have to say about mutations, the fate of humanity can be predicted; and it doesn't look good. In fact, it looks nightmarish. To give you an idea why that's the case, I would like to cite a paper written by Leslie A. Pray, Ph.D. titled *DNA Replication and Causes of Mutation*, which explains the foundation of evolution. If Dr. Pray is correct, the future of humanity is bleak. My feeling is that when Dr. Pray wrote this, she naively wasn't considering the ramifications of evolution, and the science that goes along with it, for humanity, which I have found to be the case with virtually every evo-illusion writer:

While most DNA replicates with fairly high fidelity, mistakes do happen, with polymerase enzymes sometimes inserting the wrong nucleotide or too many or too few nucleotides into a sequence. Fortunately, most of these mistakes are fixed through various DNA repair processes. Repair enzymes recognize structural imperfections between improperly paired nucleotides, cutting out the wrong ones and putting the right ones in their place. But some replication errors make it past these mechanisms, thus becoming permanent mutations. These altered nucleotide sequences can then be passed down from one cellular generation to the next, and if they occur in cells that give rise to gametes, they can even be transmitted to subsequent organismal generations. Moreover, when the genes for the DNA repair enzymes themselves become mutated, mistakes begin accumulating at a much higher rate. In eukaryotes, (animal cells) such mutations can lead to cancer... If DNA repair were perfect and no mutations ever accumulated, there would be no genetic variation—and this variation serves as the raw material for evolution. Successful organisms have thus evolved the means to repair their DNA efficiently but not too efficiently, leaving just enough genetic variability for evolution to continue.

Articles like this showing how genetic mutations form all species, organs, and biological systems usually give example diagrams of copy errors in DNA code, something like: ...**AG**TACGCTT... makes a copy error that results in: ...**GA**TACGCTT... Notice the G and A are transposed. All you need to know is DNA code is much like the 0's and 1's of digital coding, and the dots and dashes of the Morse code. Errors in any code can cause catastrophes. As the paper above states, mutations, the heart and the entire basis of evolution, occurs when DNA replicates during the formation of new gametes. (sperm and egg)

Leslie Pray doesn't realize she is describing a doomsday scenario for humanity. According to the illusion of evolution, if deleterious mutations sneak through the biochemical correcting processes, natural selection eliminates them. The animals with the deleterious mutations are killed due to predator/prey relationships, changes in climate, or competition for food or mates. The mutant animals are simply not strong enough to protect themselves from predation and natural catastrophes. The strong mutations that survived DNA's correction process, and also survived natural selection, are responsible for the formation of all living species, all organs, and all biological systems that have ever existed, and that exist today. Yes, even your brain, intelligence, consciousness, liver, heart, blood, skin... everything is formed because of these incredibly rare copy errors that are "selected for".

But evolution is a science with tunnel vision. It's meant for animals in the wild, but not for plants and people. Strong plants do not kill and consume weaker ones, and they don't do battle for mates. The major factors connected to evolution's selection and correction process are mostly *excluded* for plants. Plants are eliminated from most of the effects of natural selection, which should make one wonder if evolution is capable of originating and forming all of the millions of plant species on Earth, and their vast array of characteristics and biochemical processes. Do floods and droughts create enough selection to form flowers, weeds, artichokes, and trees, and all of their complex biological devices? Are evo-illusionists distracting their audiences by focusing almost completely on animal evolution? It's a pretty good trick, and one used by stage illusionists all the time. From the looks of it, evo-illusionists use this trick as well. "Get them to focus on animals, so they don't think about plants."

Natural selection improves the health of animal populations; but not human populations. The vast majority of mutations are deleterious for humans. According to modern science though, the bad mutations in human populations cannot be corrected through natural selection.

The fact that bad mutations that occur during the formation of human sperms and eggs aren't completely corrected, as cited by Dr. Pray, and are not removed by natural selection, is disastrous. Humans no longer exist within the predator/prey system credited by evolution's illusionists with saving beneficial mutations and removing the deleterious ones from animal populations. Humans are no longer animals in the wild.

Since humans left the predator/prey system thousands of years ago, there is no entity that removes harmful genetic mutations out of the human population. They will remain and be built on by newer deleterious mutations, generation by generation. Weaker humans can procreate nearly as frequently as do the strong healthy ones. Humans have found ways to eliminate practically all of the environmental problems that might select the

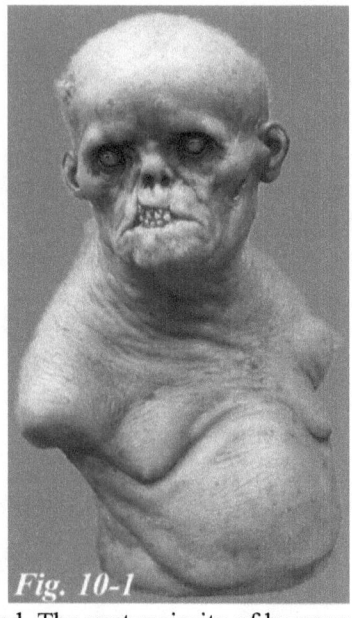

Fig. 10-1

strong and good mutations, and eliminate the bad. The vast majority of humans have efficient shelter and protection from most environmental difficulties, incredible food resources, and a plentiful supply of mates. We don't have to do battle to the death or cause severe injury to our mating opponents to win our mating partners, as is the case with animal natural selection. We have fantastic medical systems and hospitals that take care of the sick and weak. Humans have been nearly completely eliminated from the natural selection process that drives evolution. The net result is we are gradually being poisoned by our own genetic system. Mankind will become inundated with mutants like the one in Figure 10-1. At least 97 percent of mutations are deleterious or neutral. The mutations that escape cellular biochemical correction will remain in the human population, and build up, generation by generation. The future of mankind is that everyone will have cancer and severe physical limitations and illnesses due to this ongoing buildup of bad mutations. If modern biology and evolution are correct, the human population on Earth will be destroyed by its own genetic system. The only hope for humanity is if there is a currently unknown or unrecognized natural controlling and correcting mechanism that is able to repair DNA replication errors. Hopefully, there is such a controlling mechanism, and it can correct the massive numbers of mutations that have formed and will be forming in human DNA. If there is no correcting mechanism that can take over for humanity's loss of most natural selection processes, we humans will inhabit the Earth for a comparatively short time. We will have our own consciousness and intelligence, and our ability to protect ourselves from predation and most environmental upheavals, to blame for our demise. It's all downhill from here, sorry to say. Actually, there is one other circumstance that will save mankind from a horrible end. And that is, if evolution is *not* the source of all living species and biological systems, including mankind. If that's the case, then there is hope. Lot's of hope.

The year was 2001. I flew to Chicago to visit my son who was in medical school there. I was so excited, not only because I was going to visit my son, but because the Field Museum was in Chicago. They had the best fossil collection in the world. I was an avid evolution believer at the time. I was going to get to see all of the incredible new fossils that had been found in the last few decades that were sure to prove evolution without a doubt. After all, evolution was part of modern science, and modern science is so valid. It's not to be questioned. And I didn't question; until I saw that fossil collection at the Field. There was absolutely no evolution. Where was it? There were trilobite fossils over 300 million years that showed no change. And sea horses, and frogs, and nautilus, and snails, and worms, and fish, and ferns, and all sorts of other species that had existed on the Earth for hundreds of millions of years, all without change. Where was the evolution that I was so sure I was going to see? It wasn't there at all. It was like popping a giant balloon. My excitement turned to puzzlement. I walked in to the Field Museum thrilled. I left wondering. Had this particular science let me down? This is the 21st century. Are we still doing fables and myths, just like the people of all previous centuries that had questions they couldn't answer? Did we humans make up fables for the origin of organs, biological systems, and species because we don't understand how they originated? Are we no better than the ancients?

I went home from Chicago and did a great deal of research on the subject of our origins, and the origin of all living organisms. This time my research had a more critical bent. I wasn't such an easy believer as I was when I read Time-Life's series *The Emergence of Man*, and watched numerous documentaries on the origins of humans and all living species. The deeper I dug, the more I realized I had been fooled. Before my trip to the Field Museum, I revered all science. I was sure modern science was pure, honest, and true. When I found out what modern science had to say about the origin of all species and biological systems was nothing but an illusion, no different than the fables and myths of the ancients, but far better disguised. I was stunned. I began doing research on evolution. I was sincerely hoping what I found at the Field Museum was just an anomaly. Certainly sites that promoted and educated people on evolution would have what I missed at the Field: great fossil evidence proving evolution was the source of all living species and biological systems. For convenience, I collected my research on a blog (www.evoillusion.org). Evolution devotees found my blog and started arguing with some of the information I had posted. I soon realized that I needed to do an in-depth study of evolution, and review my biology, biochemistry, and physiology so I could discuss intelligently with my best and most educated blog visitors. I had more than enough classes at the University of Southern California to obtain a masters degree in biology. But I needed an update. My blog grew as I added more pages of information. As the information increased, the challenging commenters did as well. The blog gave birth to my first book *Evo-Illusion*.

Evo-Illusion covered the evolution of animal species, but I just didn't have room to discuss the evolution of man. The book would have been way over my target size of 300 pages. So I saved human evolution for my second book. My first chapter in *Evo-illusion of Man* was going to deal with genetics, and the evolution of the human brain. I looked deeper and deeper, and began analyzing the genetics of the human brain, and the amount of coding needed to construct it. The numbers simply didn't add up. I had always accepted without question that our genes hold the plans for the entire human body. They hold the plans for our brains, hearts, blood and blood vessels, ball and socket joints… everything that we're composed of. That "fact" is written in virtually every textbook on the subject, and is relayed to nearly every student in science lecture halls. Even Richard Dawkins, the Pope of modern evolution, says in his book, *The Selfish Gene*, that the DNA (genes) in each cell in the human body hold the plans for the entire human body. It's an astounding thought, but one I accepted readily when I read *The Selfish Gene*. I had no doubt Dawkins was at least right about that.

So here I was with a dilemma. I was writing about human genes and how they hold the plans for the human brain. I quickly counted way too many very basic entities in the brain that needed coding. The coding in our DNA didn't come close to matching the coding needed to assemble only the rudiments of the brain. Fruit flies have 15,000 genes. Humans have 22,000. The human brain alone is millions of times more complex than a fruit fly. Had I been fooled not only by evolution, but by genetics as well? I just didn't believe that could be possible; but was it? The more I dug, the more I thought, the more the notion that DNA holds all of the plans for the human body, or any body for any species, collapsed. Not only was evolution replete with myths that have been accepted and believed by millions of people, including me, but genetics was as well. It had to be. Because as I looked at the genetics that supported evolution, and that I had so believed and accepted as valid, I saw more and more myths and illusions that were presented by people who I considered to be highly respected scientists. I had been fooled just like most people.

I deleted the genetic and DNA discussions that I initially wrote about in my first chapters of this book, as it became too involved, and off-topic. So I'll save that for another writing. But the math is simple: the 3.2 billion bits of DNA code present in every human cell isn't enough code to construct even the most basic structure of the human brain, with its 100 billion neurons (brain cells), each with tens of thousands of dendrite interconnects. Each brain cell is like an electronic component in a circuit board. Each connection must be perfectly constructed, or the brain wouldn't work, any more than would a circuit board with dozens of bad connections. Our DNA code would have to have over 10 quadrillion bits of code just to hold the plans for the connections of the neurons in our brains. If I am right, and DNA cannot code for the human body and all of its marvelous parts, then evolution cannot be the entity that formed us all. And we are saved from the certain catastrophe that will befall us if I am wrong; and if most evolution scientists in the world are right.

The huge question is "Why"? Why would modern science and scientists try to fool so many people into believing they have the answers to so many questions that we are not close to answering? Why does modern science still have one foot stuck in the Dark Ages of science when we've come so far? Our scientific knowledge has gone light-years beyond what the scientists knew in the 19th Century. We certainly look more scientific, we are more objective, and we should be very proud. So why does modern science still need to make up illusions, myths, and fables to fool the masses into thinking they have answers that no person on Earth has? It's very hard for me to understand. My respect for modern science has floundered. I realize more than ever that modern science still requires a great deal of skepticism. I have always loved reading National Geographic and Scientific American; but now I will do it with a far more cynical eye. I now more than ever question every science documentary that I watch and every book I read. Which is a good thing. Skepticism can only improve science. Skepticism certainly makes science far more fun and interesting for me. Finding holes in scientific dogma will always be a kick.

In analyzing and thinking about why science still has such a great need to fool the masses and concoct illusions, I have come to several suppositions. Here are my explanations for this embarrassing phenomenon, born of what I have experienced personally in all of my writings, debates with people who disagree with my stance including PhD scientists, the making of videos that I have placed on the internet, and the research I have implemented for my books:

(1) We humans can't come to grips with our scientific shortcomings: Modern scientists simply can't admit they don't have all the answers. That's just the way our 100 billion neurons in our brains are wired. If we don't have the answer, we will make one up. We humans have been doing that for thousands of years. No need to stop now. There are numerous entities and phenomena that we don't have scientific explanations for. The origin of human intelligence and consciousness alone trump any attempt by any man to explain the existence of humans. There are certain questions that no person can even make up a fable about, such as why the universe exists. There isn't even a good imaginary answer to that one, so science leaves it alone. But, even though they have absolutely no idea how life or human intelligence and consciousness came

to be, they have no problem making up fables to cover that deficiency. Evo-illusionists can make fables about that, and they are just believable enough to fool the masses. But concocting an illusion about why the universe is here instead of nothing is beyond the fable-making abilities of any man. So it's just ignored.

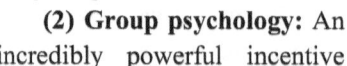
All those who think humans came from apes say aye.
"Aye!" "Aye!" "Aye" "Aye!" "Aye!"

(2) Group psychology: An incredibly powerful incentive for scientists and teachers who present and support scientific illusions is group psychology. Group psychology is far more powerful of a process than it would seem possible. If you ever wonder, just take a look at films of Germany and its citizens before and during the early parts of World War II. Germans were being taken down the most horrible road to destruction. Nazis were doing hideous things to good and great people. But group psychology took over. The German citizenry accepted all of those horrors. They were excited, happy, waving, worshipping Hitler... They had no idea that their lives were about to be destroyed. Were they unable to think rationally? Group psychology had destroyed their skepticism; their ability to think logically. Group psychology engulfed them. Their minds became slaves of the group. I am not comparing modern science with the German Nazis. But the Germany of the 1930's and 1940's is such a great example of the powerful effects group psychology has on masses of people.

Just about every scientific illusionist is tied to some university somewhere, where they lecture, and meet with other scientist and teachers. They know that if they demean the illusions that they and others present, or suggest that the illusions are just that, illusions, they will lose their jobs. They will be shunned by their peers. So rather than facing the wrath of the school administration, and the disdain of their fellow scientists and teachers, they go along. Just as did the citizens of Germany before and during World War II. In doing so they talk themselves into the validity of the illusions. If I could dig deep down inside the heads of scientists and teachers who present scientific illusions, and find out what they really believe, my bet is I could find a large percentage that are non-believers. But one goads on the others, they build each other up, and give each other awards and pats on the back. Being accepted and adored by the group feels good. So why not feel good, go along with the group, and avoid being shunned?

A perfect example is Richard Dawkins. If there were a Pope of Evolution, Richard would certainly be it. Pope Richard! Does he really believe what he so adamantly preaches? In his famous book, *The Blind Watchmaker*, considered to be the modern-day version of Darwin's *On the Origin of Species*, he writes statements on numerous pages that would seem to give a clue that if you could really get inside his head, you would find a person who doesn't really believe what he says. Here are only three examples:

Regarding the eye: *Think of all the intricately cooperating working parts (of the eye): The lens with its clear transparency, its colour correction and its correction for spherical distortion, the muscles that can instantly focus the lens in any target from a few inches to infinity; the iris diaphragm of "stopping down" mechanism, which fine-tunes the aperture of the eye continuously, like a camera with a built-in light meter, and fast special-purpose computer; the retina with its 125 million colour-coding photocells; the fine network of blood vessels that fuels every part of the machine; the even finer network of nerves-the equivalent of connecting wires and electronic chips. Hold all this fine-chiseled complexity in your mind and ask yourself whether it could have been put together by the principle of use and disuse. The answer, it seems to me, is an obvious "no".* (p. 301)

Regarding bats: *The mounting and hinging of these three bones (sonar) is exactly as a hi-fi engineer might have designed it to serve a necessary "impedance-matching" function, but that is another story... Echo-sounding by bats is just one of the thousands of examples that I could have chosen to make the point about good* **design**. *Animals give the appearance of having been* **designed** *by a theoretically sophisticated and practically ingenious physicist or engineer... Bats are like miniature spy planes, bristling with sophisticated instrumentation.* (p. 24)

Regarding the sudden appearance of species in the fossil record: *The Cambrian strata of rocks, vintage about 600 million years, are the oldest ones in which we find most of the major invertebrate groups. And we find many of them already in an advanced state of evolution, the very first time they appear. It is as though they were just planted there, without any evolutionary history.* (p. 229)

If one could dig deep into Richard's brain, my bet is you would find a person who realizes what he teaches is poppycock. Maybe one wouldn't even have to dig very deep. These statements are pretty telling. What does this statement have to say about my *population in situ idea*? As absurd as it may seem, Dawkins makes it less absurd.

(3) Money: Richard Dawkins has an estimated net worth of $135 million according to the Sunday Times in 2012. He has earned his net worth because of his book sales, science career, lectures, and television and film appearances that all involve the promotion of evolution. Not only do people make incredible amounts of money selling scientific illusions, but the jobs of university professors, scientists, and lecturers are dependent on their belief in, and support of these scientific illusions. Earning a lot of money, and keeping your job and income is certainly just about as important of an incentive for promoting illusions as real science.

(4) Power: Politicians, bureaucrats, evangelical religious charlatans, and scientific illusionists are constantly on a search to find ways they can control the minds and lives of the "masses". Some humans have an inner need to hold power over other humans; to be able to control them like a puppeteer controls his puppets. Power corrupts. That's a fact that has been proven over and over in the history of mankind. Evo-illusionist professors go in front of college audiences in huge auditoriums and lecture about evolution. They relay preposterous fables that not one student should believe, but they nearly all do. Can you imagine the feeling of power the lecturer has over the students? Talk about a drug high. This must be even better. And the students? Of course the lecturer intimidates them, and they fear that not agreeing will make them look foolish. They certainly don't want to stand out and not be accepted as part of the group. So a whole room full of students will remain silent, believe, and rarely question or protest. A student who does ask question is scoffed at and ridiculed by fellow students who are on the "right" side. The lecturer gets the high of wielding power over so many students. There certainly must be a high from fooling so many people at once.

(5) **Illusions must be concocted to cover for earlier illusions:** Illusions that are concocted when the presenter knows they aren't real, but they're sold as if they are real, are nothing but lies. Just as lies are frequently needed to support earlier lies, illusions are used to support earlier illusions. For example, genetics uses illusions to support evolution. The illusion: DNA makes body parts and holds the plans for the entire human body. Copy errors during procreation are responsible for making those parts, all species, and humans. The truth: DNA only carries code to make proteins. Illusions like this about genetics are supported by evolution. They are symbiotic. They both need the other to survive and be accepted by audiences. They are intimately tied together. One illusion builds on the next and the next. Once an illusion is presented and accepted, new questions arise which require the creation of another new illusion. Illusions pile up, one on top of the next, until a mountain of illusions has formed. The higher the mountain of illusions, the more difficult it is to challenge any one illusion.

(6) **Some scientists are just downright gullible:** This is probably the most basic of reasons that answers the question about why scientific illusionists do what they do. It's entirely possible that when they were young and in school, they were sold on evolution to such a degree that their brains are perfectly and permanently wired into being believers in the illusions that they were shown. They have been 100% indoctrinated, and they are going to pass on the indoctrination to their students and their audiences. They are 100% sure they

are right. There is no other possibility, including the possibility that humanity isn't close to figuring out the puzzles that nature has presented to us. They will continue doing their evo-illusions and being deep believers in those illusions for the rest of their lives. They will unfortunately sell these illusions to their audiences, who will believe in, group form.

(6) **Do battle with religion:** There is a bloodless war going on right now between atheism and religious creationism. Evo-illusion gives atheism a backbone; a *raison d'etre*. For without the illusion of evolution, how would atheists say all living creatures came into existence? So atheism needs evolution. Evolution actually doesn't need atheism. Many religious people think evolution is the way God brought man and all of living nature into being. They think evolution is a tool of God. Ask the Catholic Pope. Just as religions are supported by many parables and illusions, so too are atheism and evolution. They both try to answer questions that mankind is unable to answer.

I'm sure there are many more reasons evo-illusionists and other scientific illusionists have for promoting their illusions. Every case is different, of course. But there is an overall pall over modern science that ties it to the science of the Dark Ages. We in the 21st century simply have a different view than did people of far earlier times. Will the illusions that I have enumerated and described in this book ever be dispatched, and replaced with good objective science? As long as the answers to major scientific questions remain unanswered, and are unanswerable, I fear that these illusions will have a very long life. We humans simply cannot accept that we don't have all the answers. If enough people want to believe rather than use their natural intelligence and abilities to be skeptical, and if scientific illusionists keep promoting their illusions, and coming up with new ones, which they will, the illusions of modern science will remain as a malignancy that damages the great and wonderful scientific discoveries that the world of modern science and diligent honest scientists have revealed. We will remain for many years in the very awkward position of scientifically having one foot still stuck in the Dark Ages, and one in the 21st century. Out of all of my research and writing, I think what really astounds me the most is that so many revered and respected institutions, such as the National Aeronautics and Space Administration, the Smithsonian Institution, National Geographic, Scientific American, Public Broadcasting Services, British Broadcasting Corporation, and NOVA absorb and accept any information and so-called evidence regarding the subjects in this book without the slightest bit of skepticism and due diligence. That I will never understand.

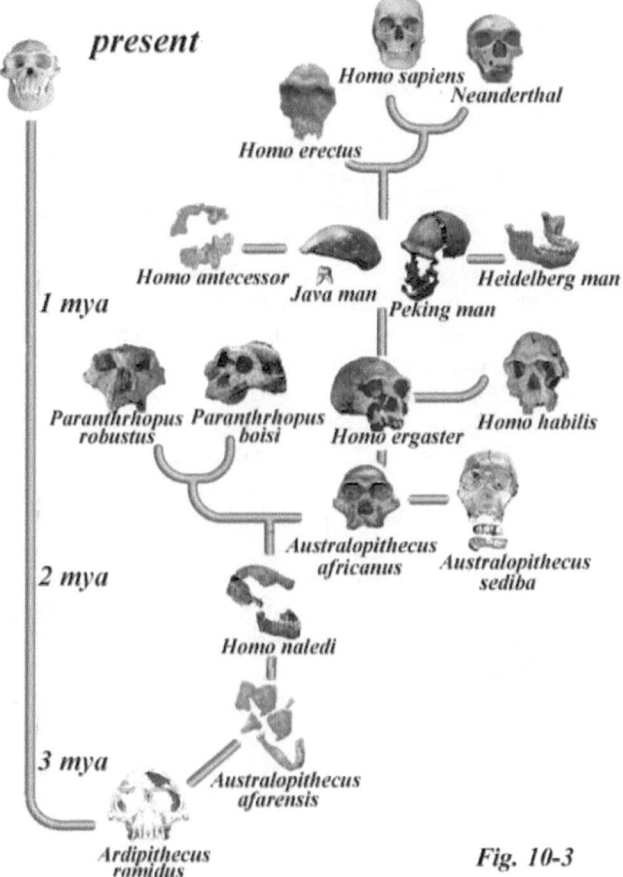

present

Homo sapiens
Neanderthal

Homo erectus

1 mya

Homo antecessor
Java man
Peking man
Heidelberg man

Paranthrhopus robustus
Paranthrhopus boisi
Homo ergaster
Homo habilis

Australopithecus africanus
Australopithecus sediba

2 mya

Homo naledi

3 mya

Australopithecus afarensis

Ardipithecus ramidus

Fig. 10-3

So what are the answers to the great *Puzzles* I have enumerated in this book and in *Evo-illusion*? The answers are simple. Just as we don't know why the universe exists instead of nothing, we don't know the source of life, biological systems, plant species, animal species, and human beings. We aren't close to answering these questions. As long as the myths and fables of modern science remain in place, we never will. Probably we will never know the answers to these questions anyway.

On the first page of Chapter 1 I placed a tree of life for humans, as portrayed by evo-illusionists. It shows how the *models* made by evo-artists, not formerly living creatures, evolved into Homo sapiens. I will finish this book with a "real" tree of life (Figure 10-3) that shows the incredibly pathetic skulls the models on the first page were supposedly modeled after. This is a true tree of life, but one you will never see presented by any evo-illusionist. It would kill the illusion. In reality, there is no "tree of life" for humans or any animal species. We didn't arise from one common ancestor as evo-illusion says. (See Chapter 11in *Evo-illusion*.) But I will use evo-illusion's own tree as a format to demonstrate their pathetic evidence. Remember, the unbelievably piecemeal group of bones on this tree is the entire evidence that humans evolved from apes over the last several million years; a huge conclusion from a pittance of evidence. Questions arise. Why are Mr. Robustus and Mr. Boisei even on this tree? Both are obvious apes. Of course, to promote the illusion that the tree of life is crowded with evidence. Ms. Afarensis, and Mr. naledi, in fact all hominids, are put together by the imagination of modelers. The four geographic hominids at 800,000 years, Mr. Antecessor (Spain), Java man, Peking man, and Heidelberg man are bits of bone that were turned into hominids by evo-artists. They are there because the countries that they were found in want their own chunk of the human family tree. They want their own hominid. It's beyond obvious that no ape 800,000 years ago would have trekked to any of these geographic locations from Africa. Every one of the guys at the 800,000-year mark should be tossed. The entire family tree is a disaster. Seeing all of these guys together, and realizing what the evo-illusion of man is composed of, it's really pretty disappointing. But what an incredible illusion!

So what is that answer to the *Puzzle* of how humans came to be on this Earth? And, how did they spread themselves all over the globe, with no significant or provable pathway for doing so? Did intelligent beings like humans form on any other planet in the universe? Is there life of any kind anywhere in the universe? It's clearly obvious that there is currently no objective and scientific answer to any of these questions. As good as human technology and knowledge have become, and as smart as we humans are, we don't have enough technology and smarts to even begin to answer these questions. So we do what scientists of the Dark Ages did. We make up fables. The fables are verified by agreement. They become real answers for many believers. Which means we will never know any of the solutions to the *Puzzles* that we so desperately want solved. And that is sad.

References:

Chapter 1: Hominid Lesson

1. Smithsonian: Bone vs. Stone: How To Tell The Difference: http://www.smithsonianmag.com/science-nature/bone-vs-stone-how-to-tell-the-difference-62895060/#82yihFdETLczEibs.99
2. Endangered Species International, http://www.endangeredspeciesinternational.org/overview.html
3. PBS, Evolution, Deep Time, http://www.pbs.org/wgbh/evolution/change/deeptime/low_bandwidth.html
4. NOVA, Becoming Human, The Birth of Humanity; 2009
5. Diseases and Conditions: Microcephaly, Mayo Clinic: http://www.mayoclinic.org/diseases-conditions/microcephaly/basics/definition/con-20034823
6. National Institute of Neurological Disorders and Stroke, Microcephaly Information Page http://www.ninds.nih.gov/disorders/microcephaly/microcephaly.htm
7. http://uts.cc.utexas.edu/~bramblet/ant301/seven.html
8. Wikipedia, Brain Size, http://en.wikipedia.org/wiki/Brain_size
9. NECSI: Gradualism and Punctuated Equilibrium: http://necsi.edu/projects/evolution/evolution/grad+punct/evolution_grad+punct.html
10. The Descent of Man : http://darwin-online.org.uk/EditorialIntroductions/Freeman_TheDescentofMan.html
11. Smithsonian Museum of Natural History, What Does it Mean to be Human: Homo habilis, http://humanorigins.si.edu/evidence/human-fossils/species/homo-habilis
12. Science AAAS, Paleoanthropology: Tracing the Identity of the First Toolmakers, by Ann Gibbons April 4, 1997
13. http://www.sciencemag.org/content/276/5309/32.summary
14. The Descent of Man by Charles Darwin: http://human-nature.com/darwin/descent/chap3.ht
15. The Blind Watchmaker, by Richard Dawkins, W.W. Norton & Company, New York, London, 1987, p. 228

Chapter 2: Piltdown

1. Neanderthals: Facts About Our Extinct Relatives, Jessie Szalay, March 19, 2013 http://www.livescience.com/28036-neanderthals-facts-about-our-extinct-human-relatives.html
2. Top 10 Misconceptions About Neanderthals http://listverse.com/2009/06/16/top-10-misconceptions-about-neanderthals/
3. Smithsonian Museum of Natural History http://humanorigins.si.edu/evidence/human-fossils/species/homo-neanderthalensis
4. Last of the Neanderthals, National Geographic, Stephen S. Hall, October 2008, http://ngm.nationalgeographic.com/2008/10/neanderthals/hall-text
5. Why Don't We Call Them Cro-Magnon Any More, bye Kris Hirst, About Education http://archaeology.about.com/od/earlymansites/a/cro_magnon.htm
6. Fagan, B.M. (1996). The Oxford Companion to Archaeology. Oxford, UK: Oxford University Press. p. 864. ISBN 978-0-19-507618-9. "The Cro-Magnons are identified with Homo sapiens sapiens of modern form, in the time range ca. 35,000-10,000 b.p. [...]
7. Wilford, John Noble (November 2, 2011). "Fossil Teeth Put Humans in Europe Earlier Than Thought". The New York Times. Retrieved June 8, 2012.
8. "Cro-Magnon (anthropology) - Britannica Online Encyclopedia". Britannica.com. Retrieved 2011-10-05.
9. Smithsonian National Museum of Natural History: What Does it Mean to be Human; http://humanorigins.si.edu/evidence/human-fossils/fossils/cro-magnon-1
10. Piltdown Man, Kenneth F. Oakley and J. S. Weiner, American Scientist, October 1955, British Museum of Natural History, and University of Oxford

11. Natural History Museum, Piltdown Timeline
    http://www.nhm.ac.uk/nature-online/science-of-natural-history/the-scientific-process/piltdown-man-hoax/timeline/index.html
12. Piltdown Man, Kenneth F. Oakley and J. S. Weiner, British Museum of Natural History, American Scientist, Oct. 1955
    http://www.clarku.edu/~piltdown/map_gen_hist_surveys/piltman_oaklywiener.html
13. The Piltdown Men, Ronald Millar, St. Martin's Press, New York, Library of Congress No. 72-94380, 1972
14. Piltdown Man, The Bogus Bones Caper, by Richard Harter, 1996-7
    http://www.talkorigins.org/faqs/piltdown.html
15. Piltdown Man is Revealed as a Fake, 1953, A Science Odyssey
    http://www.pbs.org/wgbh/aso/databank/entries/do53pi.html
16. Encyclopedia Britannica, Piltdown Man, July 17, 2014
    http://www.britannica.com/EBchecked/topic/460690/Piltdown-man
17. NOVA Science, The Boldest Hoax, January 11, 2005
    http://www.pbs.org/wgbh/NOVA/hoax/
18. Russell in his book Piltdown Man: The Secret Life of Charles Dawson.
19. Piltdown Man: Archaeology's Greatest Hoax by Robin McKie
    http://www.theguardian.com/science/2012/feb/05/piltdown-man-archaeologys-greatest-hoax
20. letter from Osborn to Cook, 1922b, p. 2: http://www.talkorigins.org/faqs/homs/wolfmellett.html
21. Science Magazine, Vol. 55 no. 1427, pp. 463-465, Hesperopithecus, The First Anthropoid Primate Found in America, May 5, 1922 http://www.sciencemag.org/content/55/1427/463.short
22. National Center for Science Education, The Role of Nebraska Man in the Creation-Evolution Debate, by John Wolf and James S. Mellett, Vol 5, #2, Pages 31-43, 1985
    http://ncse.com/cej/5/2/role-nebraska-man-creation-evolution-debate
23. American Museum Novitates,, April 25, 1922: Hesperopithecus, The First Anthropoid Primate Found in America,
    http://digitallibrary.amnh.org/bitstream/handle/2246/3251/N0037.pdf;jsessionid=4BFE6C6454A5C8A76DEDF9860CDFA0D9?sequence=1
24. The Role of Nebraska Man in the Creation/Evolution Debate
    http://www.talkorigins.org/faqs/homs/wolfmellett.html
25. Nebraska Man: http://en.wikipedia.org/wiki/Nebraska_Man
26. About Education, Nebraska Man, by Heather Scoville, evolution expert;
    http://evolution.about.com/od/controversy/a/Nebraska-Man.htm

Chapter 3: Java man and Peking man

1. Encyclopedia Britannica: Java man http://www.britannica.com/EBchecked/topic/301721/Java-man
2. The Australian Museum: The First Modern Humans in Southeast Asia:
   http://australianmuseum.net.au/the-first-modern-humans-in-southeast-asia
3. Smithsonian National History of Natural History, What Does it Mean to be Human: Trinil 2, Indonesia's Java man: http://humanorigins.si.edu/evidence/human-fossils/fossils/trinil-2
4. *Foley/Milton Debate: http://www.talkorigins.org/faqs/homs/milton5a1.html*
5. *Was Java man a Gibbon? http://www.talkorigins.org/faqs/homs/gibbon.html*
6. Gould S.J. Men of the thirty-third division. pp. 124-37. New York: W.W.Norton., 1933
7. http://www.talkorigins.org/faqs/homs/java.html
8. Great Archeology: Java man, http://www.greatarchaeology.com/java_man.htm
9. http://en.wikipedia.org/wiki/Java_Man
10. Was Java Man a Gibbon?: http://www.talkorigins.org/faqs/homs/gibbon.html
11. Smithsonian Museum of Natural History: http://humanorigins.si.edu/evidence/human-fossils/species/homo-heidelbergensis

12. http://www.talkorigins.org/faqs/homs/mauer.html
13. Australian Museum, Homo heidelbergensis: http://australianmuseum.net.au/homo-heidelbergensis
14. Smithsonian National Museum of Natural History, What It Means to be Human: Homo heidelbergensis
15. http://humanorigins.si.edu/evidence/human-fossils/species/homo-heidelbergensis
16. Smithsonian National Museum of Natural History, What It Means to be Human: Oldest Wooden Spear, http://humanorigins.si.edu/evidence/behavior/oldest-wooden-spear
17. http://en.wikipedia.org/wiki/Sch%C3%B6ningen_Spears
18. Australian Museum, Nature, Culture, Discover: Homo heidelbergensis http://australianmuseum.net.au/Homo-heidelbergensis#sthash.sdli0aCU.dpuf
19. Science Daily: Homo heidelbergensis http://www.sciencedaily.com/articles/h/homo_heidelbergensis.htm
20. Discovery: Heidelberg man Links Humans and Neanderthals May 24, Jennifer Viegas http://news.discovery.com/history/archaeology/humans-neanderthals-heidelberg-man-110504.htm
21. Biotechnology and Life Sciences in Baden-Württemberg: Homo heidelbergensis http://www.bio-pro.de/magazin/thema/00145/index.html?lang=en&artikelid=/artikel/08962/index.html
22. The Rockefeller Foundation and the Excavation of Peking Man By Chris Manias, Lecturer, University of Manchester, Manchester UK 2012 http://www.rockarch.org/publications/resrep/manias.pdf
23. Talk Origins, The Lost Peking Man Skeletons: *http://www.talkorigins.org/faqs/homs/lostskels.html#initial*
24. Natural History Museum http://www.nhm.ac.uk/nature-online/life/human-origins/early-human-family/index.html
25. Encyclopedia Britannica: Peking Man: http://www.britannica.com/EBchecked/topic/448989/Peking-man
26. Talk Origins: Peking Man: http://www.talkorigins.org/faqs/homs/peking.html
27. Peking Man World Heritage Site: http://www.unesco.org/ext/field/beijing/whc/pkm-site.htm
28. Australian Museum: The First Modern Humans in Southeast Asia: http://australianmuseum.net.au/the-first-modern-humans-in-southeast-asia
29. Evidence shows that a venerable cave was neither hearth nor home. By Noel T. Boaz and Russell L. Ciochon, http://www.uiowa.edu/~bioanth/courses/Peking1.htmNew
30. The Renewed Search for Peking Man China Heritage Newsletter September, 2005 The Australian National University http://www.chinaheritagequarterly.org/articles.php?searchterm=003_pekingman.inc&issue=003
31. The Bizarre Disappearance of the Peking Man Fossils, by Ester Inglis-Arkell, February 1, 2013: http://io9.com/5980372/the-bizarre-disappearance-of-the-peking-man-fossil
32. Crime Library: Dragon Bones: The Mystery of Peking Man, by Rachael Bell, http://www.crimelibrary.com/criminal_mind/forensics/peking_man/6.html
33. Unique Canine Tooth from Peking Man Found, May 11, 2011 http://phys.org/news/2011-05-unique-canine-tooth-peking.html
34. The Peking Man World Heritage Site at Zhoukoudian http://www.unesco.org/ext/field/beijing/whc/pkm-site.htm
35. Travel China Guide, Peking Man Site at Zhoukoudian: http://www.travelchinaguide.com/attraction/beijing/pekingman.htm

Chapter 4: Modeling

1. PBS Library, The Scopes Trial:
   http://www.pbs.org/wgbh/evolution/library/08/2/l_082_01.html
2. Smithsonian Institution Archives: The Scopes Trial
   http://www.siarchives.si.edu/research/scopes.html
3. Vasa Museum: From Wreck to State of the Art http://www.vasamuseet.se/en/
4. Evolution: The Human Story, by Alice Roberts, Dorling Kindersley, New York, 2011
5. Kennis and Kennis Reconstructions: http://kenniskennis.com/site/Home/
6. National Geographic online, November 4, 2012 http://juanvelascoblog.com/2012/11/04/artist-profile-the-incredible-kennis-brothers/
7. NPR News: Earliest Human Footprints Found in Kenya:
   http://www.npr.org/templates/s/s.php?sId=101191786
8. Live Science, The Shoe Fits: 1.5 Million Year Old http://www.livescience.com/5344-shoe-fits-1-5-million-year-human-footprints.html
9. The Guardian: Earliest Human Footprints Found in Kenya
   http://www.theguardian.com/world/2009/feb/27/earliest-human-footprint-found
10. Scientific American, The Human Pedigree, by Kate Wong, Jan, 2009, pp. 61-63
11. Australian Museum of Natural History: Homo Ergaster http://australianmuseum.net.au/homo-ergaster
12. Early Human Evolution: A Survey of the Biological and Cultural Evolution of Homo habilis and Homo Erectus: http://anthro.palomar.edu/homo/homo_2.htm
13. Archaeologyinfo.com Where Human Evolution and Archaeology Intersect:
   http://archaeologyinfo.com/homo-ergaster/
14. Science & Nature: Prehistoric Life, BBC
   http://www.bbc.co.uk/sn/prehistoric_life/human/human_evolution/leaving_home1.shtml
15. Discovery of prehistoric stone tools in Spain, Aug. 3, 2014
   http://www.bradshawfoundation.com/news/origins.php?id=Discovery-of-prehistoric-stone-tools-in-Spain
16. Homo antecessor, 5/14/2000 http://www.modernhumanorigins.net/antecessor.html
17. National Geographic, Jane J. Lee, Feb. 7, 2014
   http://news.nationalgeographic.com/news/2014/02/140207-ancient-human-footprints-outside-africa-england-anthropology-science/
18. Daily Tech: by Jason Mick, Feb. 10, 2014 Fossilized Footprints in British Sand Reveal Recent Relative of Man
   http://www.dailytech.com/Fossilized+Footprints+in+British+Sand+Reveal+Recent+Relative+of+Man/article34297c.htm
19. The Independent: Meet the million-year-olds: Human footprints found in Britain are the oldest ever seen outside of Africa; David Keys, Feb 7, 2014
   http://www.independent.co.uk/news/science/archaeology/news/millionyearold-norfolk-footprints-just-who-were-homo-antecessor-and-how-did-they-arrive-in-britain-9115354.html
20. Natural His Museum: We were here: earliest humans leave prints on Norfolk beach, Feb. 7, 2014http://www.nhm.ac.uk/nature-online/life/human-origins/early-human-family/homo-antecessor/index.html

21. CBS News: Scientists Discover 800,000 year-old "pioneer man" footprints in England, or Homo antecessor. http://www.cbsnews.com/news/scientists-discover-800000-year-old-pioneer-man-footprints-in-england/

22. Earliest human footprints outside Africa discovered in Norfolk: 800,000-year-old imprints 're-write our understanding of his' By Ellie Zolfagharifard and Victoria Woollaston and Sarah Griffiths Feb. 7, 2014 http://www.dailymail.co.uk/sciencetech/article-2553798/Earliest-human-footprints-outside-Africa-discovered-NORFOLK-800-000-year-old-imprints-shed-light-movement-ancient-ancestors.html#ixzz3IvPwYaDL

23. Oldest human footprints outside Africa found in UK: New Scientist 18:52 07 February 2014 by Andy Coghlan http://www.newscientist.com/article/dn25022-oldest-human-footprints-outside-africa-found-in-uk.html#.VGZ7GvTF8mY

24. Nov.13, 2014 Headlines & Global News: Earliest human footprints outside Africa discovered in NORFOLK: 800,000-year-old imprints 're-write our understanding of his http://www.hngn.com/articles/15/20140207/800-000-year-old-footprints-suggest-homo-antecessor-children-and-adults-were-headed-south.htm

25. Smithsonian National Museum of Natural His, What Does It Mean To Be Human: http://humanorigins.si.edu/evidence/human-fossils/species/australopithecus-afarensis

26. Archeology Info: Australopithecus afarensis http://archaeologyinfo.com/australopithecus-afarensis/

27. Jones, S., Martin, R. & Pilbeam, D., ed. (1994). *The Cambridge Encyclopedia of Human Evolution*. Cambridge: Cambridge University Press

28. McPherron, Shannon P.; Zeresenay Alemseged; Curtis W. Marean; Jonathan G. Wynn; Denne Reed; Denis Geraads; Rene Bobe; Hamdallah A. Bearat (2010). "Evidence for stone-tool-assisted consumption of animal tissues before 3.39 million years ago at Dikika, Ethiopia". *Nature* 466 (7308): 857–860

29. Evolution: The Human Story, by Alice Roberts, Dorling Kindersley, New York, 2011, p. 117

30. McPherron, Shannon P.; Zeresenay Alemseged; Curtis W. Marean; Jonathan G. Wynn; Denne Reed; Denis Geraads; Rene Bobe; Hamdallah A. Bearat (2010). "Evidence for stone-tool-assisted consumption of animal tissues before 3.39 million years ago at Dikika, Ethiopia". *Nature* 466 (7308): 857–860

31. Baby Lucy—The World's Oldest Child—Found By Fossil Hunters, by James Owen, for National Geographic News, September 20, 2006: http://news.nationalgeographic.com/news/2006/09/060920-lucys-baby.html

32. National Geographic: Fast Facts on an Early Human Ancestor, Sept 20, 2006 http://news.nationalgeographic.com/news/2006/09/060920-lucy.html

33. PBS: A Science Odyssey, People and Discoveries, Johanson finds 3.2 Million Year Old Lucy: http://news.nationalgeographic.com/news/2006/09/060920-lucy.html

34. Famed Lucy Fossil Discovered in Ethiopia 40 Years Ago, His in the Headlines, November 24, 2014, by Andrew Evans http://www.his.com/news/famed-lucy-fossils-discovered-in-ethiopia-40-years-ago

35. NOVA, In Search of Human Origins, funded by Merck and Prudential, June 3, 1997

36. Smithsonian National Museum of Natural His, Human Origins, http://humanorigins.si.edu/evidence/human-family-tree

37. Scientific American January 9, 2009, The Human Pedigree, by Kate Wong, p. 61-3

38. Lucy: The Beginnings of Humankind, Donald C. Johanson, and Edey Maitland, New York: Simon & Schuster, 1981, pp. 257-8, 277
39. http://en.wikipedia.org/wiki/Homo_habilis
40. PBS Evolution Home: Origins of Humankind: Orhttp://www.pbs.org/wgbh/evolution/humans/humankind/j.html
41. Archeology Information: Where human evolution and archeology intersect: Homo habilis http://archaeologyinfo.com/knm-er-1813/
42. Australian Museum, Nature Discover, Homo habilis: http://australianmuseum.net.au/Homo-habilis#sthash.spqaacKp.dpuf
43. Smithsonian National Museum of Natural His, What Does it Mean to be Human: Australopithecus africanus http://humanorigins.si.edu/evidence/human-fossils/species/australopithecus-africanus
44. Archeology Info: Australopithecus africanus: http://archaeologyinfo.com/australopithecus-africanus/

Chapter 5: New Hominids
1. The National Museum of Natural History, What Does it Mean to be Human: *Paranthropus boisei* http://humanorigins.si.edu/evidence/human-fossils/species/paranthropus-boisei
2. Science News: Paranthropus boisei: 1.34-Million-Year-Old Hominin Found in Tanzania: http://www.sci-news.com/othersciences/anthropology/science-paranthropus-boisei-hominin-tanzania-01603.html
3. Archeology Info: Ardipithecus ramidus: http://archaeologyinfo.com/ardipithecus-ramidus/
4. The Smithsonian National Museum of Natural History: What Does it Mean to be Human: Ardipithecus ramidus http://humanorigins.si.edu/evidence/human-fossils/species/ardipithecus-ramidus
5. Ardipithecus ramidus, by Eugene M. McCarthy, PhD: http://www.macroevolution.net/ardipithecus-ramidus.html
6. 6. http://www.nydailynews.com/news/world/student-finds-2-8-million-year-old-human-jawbone-ethiopia-article-1.2141719 Student finds 2.8 million-year-old human genus jawbone in Ethiopia By Alejandro Alba New York Daily News Sunday, March 8, 2015,
7. CNN: Oldest Known Jawbone From Human Genus Found in Ethiopia, by Laura Smith-Spark, March 5, 2015 http://www.cnn.com/2015/03/05/africa/ethiopia-ancient-jawbone-discovery/
8. Indi-Uni: Archeologie & Anthropology; Australopithecus sediba: Part Ape, Part Human http://www.archeolog-home.com/pages/content/malapa-afrique-du-sud-australopithecus-sediba-part-ape-part-human.html
9. *The Human Pedigree: A Timeline of Hominid Evolution*: Some 180 years after unearthing the first human fossil, paleontologists have amassed a formidable record of our forebears By Kate Wong, Scientific American, January 2009
10. Part Ape, Part Human: A new ancestor emerges from the richest collection of fossil skeletons ever found. By Josh Fischman National Geographic, August 11, 2011 http://ngm.nationalgeographic.com/2011/08/malapa-fossils/fischman-text
11. Express: New Species of Human-the Homo Naledi-Discovered by scientists in cave in South Africa, by Emily Fox September 10, 2015: http://www.express.co.uk/news/science/604208/Scientists-discover-new-species-of-human-the-homo-naledi-in-a-cave-in-South-Africa

12. National Geographic: New Human Ancestor Elicits Awe and Many Questions, by Jamie Shreeve, Sept 10, 2015: http://news.nationalgeographic.com/2015/09/150910-homo-naledi-human-ancestor-species-reaction-science/
13. The Telegraph: First Look: How Scientists Discovered our new Human Ancestor Homo Naledi, September 10, 2015 http://www.telegraph.co.uk/news/science/science-news/11855516/First-look-How-scientists-discovered-our-new-human-ancestor-Homo-naledi.html
14. National Geographic: Mystery Lingers Over Ritual Behavior of New Human Ancestor, By Nadia Drake, September 15, 2015 http://news.nationalgeographic.com/2015/09/150915-humans-death-burial-anthropology-Homo-naledi/
15. PBS: Why Did *Homo naledi* Bury Its Dead? By Nadia Drake on Tue, 15 Sep 2015
16. http://www.pbs.org/wgbh/NOVA/next/evolution/why-did-homo-naledi-bury-its-dead/
17. Meet the Man Who Gives Ancient Human Ancestors Their Faces, Paleo artist John Gurche created Homo naledi's face by making hundreds of minute anatomical calculations. http://news.nationalgeographic.com/2015/09/150914-homo-naledi-ancient-human-face/
18. The National Museum of Natural History, What Does it Mean to be Human: *Paranthropus robustus* http://humanorigins.si.edu/evidence/human-fossils/species/paranthropus-robustus

Chapter 6: Population

1. Population Education: https://www.populationeducation.org/content/what-doubling-time-and-how-it-calculated1.
2. http://geography.about.com/od/obtainpopulationdata/a/worldpopulation.htm
3. http://www.querycat.com/faq/fb6814950d4f...
4. http://www.polarbearsinternational.org/about-polar-bears/faqs#q10
5. http://www.polarbearsinternational.org/about-polar-bears/faqs#q3
6. Evolution of the Brown Bear: https://prezi.com/n0osgjcubmhz/evolution-of-the-brown-bear/
7. http://www.answerbag.com/q_view/487344
8. http://www.polarbearsinternational.org/about-polar-bears/essentials/evolution
9. History Channel Online: Black Death http://www.history.com/topics/black-death
10. The Middle Ages.net: Black Death: Bubonic Plague http://www.themiddleages.net/plague.html
11. Population Reference Bureau:http://www.prb.org/Publications/Lesson-Plans/HumanPopulation/PopulationGrowth.aspx
12. How Many Times Has the Human Population Doubled? Warren M. Hern University of Colorado http://www.drhern.com/pdfs/doubling.pdf
13. Colorado Alliance for Immigration Reform: Exponential growth, doubling time, and the Rule of 70 http://www.cairco.org/reference/exponential-growth-doubling-time-rule-70
14. Population Growth over Human History http://www.globalchange.umich.edu/globalchange2/current/lectures/human_pop/human_pop.html

Chapter 7: Migration

1. Has the Sahara Always Been a Desert?: Softpedia http://news.softpedia.com/news/Has-Sahara-Always-Been-a-Desert-47128.sht
2. The Conservation Institute: Nine Interesting Facts About the Sahara Desert http://www.conservationinstitute.org/interesting-sahara-desert-facts
3. How Deep is the Sand in the Sahara Desert http://piecubed.co.uk/sand-facts/
4. Smithsonian: The Sahara is Millions of Years Older than Thought. By Sarah Zielinski, Sept 17, 2014 http://www.smithsonianmag.com/science/sahara-millions-years-older-thought-180952735/?no-ist
5. The Australian Museum of Natural History: The First Migrations Out of Africa: http://australianmuseum.net.au/the-first-migrations-out-of-africa
6. The Oldest Homo sapiens: Fossils Push Human Emergence Back To 195,000 Years Ago http://archive.unews.utah.edu/news_releases/the-oldest-homo-sapiens/

7. Anatomically Modern human: Wikipedia:
    https://en.wikipedia.org/wiki/Anatomically_modern_human
8. The Genographic Project: The Human Journey: Migration Routs:
    https://genographic.nationalgeographic.com/human-journey/
9. The Bradshaw Foundation: http://bradshawfoundation.com/journey/
10. The Great Human Migration: Why humans left their African homeland 80,000 years ago to
    colonize the world
    http://www.smithsonianmag.com/history/the-great-human-migration-13561/?no-
    ist=&webSyncID=13d549be-a1db-2813-8d64-873543720c49&page=5
11. The Migration of Man: A Genetic Odyssey: http://press.princeton.edu/chapters/i7442.html
12. Early Human Migrations: https://en.wikipedia.org/wiki/Early_human_migrations
13. http://www.eurekalert.org/pub_releases/2004-11/uou-hrm111204.phpHow   running   made   us
    human *Endurance Running Let Us Evolve To Look The Way We Do;* University Of Utah;
    November 17, 2004
14. Why Humans and Their Fur Parted Ways By Nicholas Wade Published: August 19, 2003
    http://www.nytimes.com/2003/08/19/science/why-humans-and-their-fur-parted-
    ways.html?pagewanted=all
15. Lewis and Clark Expedition: http://www.u-s-history.com/pages/h475.html
16. Hawaii's   Hokulea   Canoe   Tells   Story   of   Polynesian   Voyage   by   Lee   Foster
    http://www.fostertravel.com/hawaiis-hokulea-canoe-tells-story-of-polynesian-voyage/comment-
    page-1/#comment-25638
17. The Princeton University Press: An Interview with Spcencer Wells
    http://press.princeton.edu/chapters/i7442.html

Chapter 8: Rocket ship XM
1. http://www.biography.com/people/nicolaus-copernicus-9256984#synopsis
2. http://en.wikipedia.org/wiki/Copernican_heliocentrism
3.   http://en.wikipedia.org/wiki/Galileo_Galilei
4. http://www.biography.com/people/isaac-newton-9422656#synopsis
5. http://www.speed-light.info/measure/roemer.htm
6. http://cosmology.carnegiescience.edu/timeline/1861 *1861: William and Margaret Huggins
    Show Stars are Suns*
7. http://www.phys-astro.sonoma.edu/BruceMedalists/Schiaparelli/
8. http://www.space.com/19774-percival-lowell-biography.html, Percival Lowell Biography,
    by Nola Taylor Redd, space. com Contributor, Feb. 13, 2013
9. http://cosmology.carnegiescience.edu/timeline/1949

Chapter 9: Is Anyone Out There?
1. Time 06/19/2015: The Search for Life in the Universe: Is Anybody Out There?
2. Wikipedia, Search for Extraterrestrial Intelligence
    https://en.wikipedia.org/wiki/Search_for_extraterrestrial_intelligence

3. BBC News: Scientists In US Are Urged To Seek Contact With Aliens, Feb 12, 2015, by
    Pallab Ghosh, science correspondent http://www.bbc.com/news/science-environment-
    31442952
4. What Is: Search for Extraterrestrial Intelligence (SETI)
    http://whatis.techtarget.com/definition/Search-for-Extraterrestrial-Intelligence-SETI
5. http://saturn.jpl.nasa.gov/science/moons/enceladus/

6. NOVA, Finding Life Beyond Earth, Are We Alone, 2011, PBS
7. http://starchild.gsfc.nasa.gov/docs/StarChild/space_level2/mars.html
8. http://historicspacecraft.com/Probes_Mars.html
9. http://mars.jpl.nasa.gov/msl/multimedia/images/?ImageID=3454
10. http://www.universetoday.com/15509/what-is-the-largest-moon-in-the-solar-system/
11. http://www.universetoday.com/19677/diameter-of-the-moon/
12. http://solarsystem.nasa.gov/planets/profile.cfm?Object=Sat_Titan
13. http://solarsystem.nasa.gov/europa/overview.cfm
14. http://www.foxnews.com/science/2014/08/25/jupiter-icy-moon-europa-best-bet-for-alien-life/ Jupiter's icy moon Europa: Best bet for alien life? By Nola Taylor Redd, SPACE.com Contributor Published August 25, 2014
15. http://solarsystem.nasa.gov/planets/profile.cfm?Object=Jup_Europa
16. http://www.space.com/44-venus-second-planet-from-the-sun-brightest-planet-in-solar-system.html
17. http://solarsystem.nasa.gov/planets/profile.cfm?Object=Venus
18. http://www.space.com/15716-alien-life-search-solar-system.html (Six Most Likely Places for Alien Life in the Solar System)
19. A Tiny Wobble Reveals A Massive Planet By Phil Plait | May 29, 2009 7:00 am http://blogs.discovermagazine.com/badastronomy/2009/05/29/a-tiny-wobble-reveals-a-massive-planet/#.Vb2yOGRVhBc
20. NASA: Five Ways to Find a Planet http://planetquest.jpl.nasa.gov/system/interactable/11/index.html
21. http://abyss.uoregon.edu/~js/ast122/lectures/lec10.html
22. www.space.com/17137-how-hot-is-the-sun.html
23. Astronomical Society, AAS Meeting #221, #333.02 Publication Date:01/2013 - http://adsabs.harvard.edu//abs/2013AAS...22133302G
24. "Magnetospheres of Earth-Like Exoplanets in Close-In Habitable Zones" Astrobiology 7 (1): 167–184: http://online.liebertpub.com/toc/ast/7/1
25. Alpert, Mark. "Red Star Rising: Scientific American". Sciam.com. Retrieved 2013-01-19.
26. Zuluaga, J. I., Cuartas P. A., Hoyos, J. H. (2012): "Evolution of magnetic protection in potentially habitable terrestrial planets", ApJ (submitted), arXiv:1204.0275
27. Cain, Fraser; Gay, Pamela (2007). "Astronomy Cast episode 40: American Astronomical Society Meeting, May 2007". Universe Today. Retrieved 2007-06-17.
28. http://www.space.com/23772-red-dwarf-stars.html
29. http://www.space.com/16673-gliese-581g-habitable-planet-existence.html
30. NASA discovers Earth's bigger, older cousin, Kepler 452b
31. https://www.youtube.com/watch?v=MlikCebQSlY (Carl Sagan - Cosmos - Drake Equation)
32. Drake, F.; Sobel, D. (1992). Is Anyone Out There? The Scientific Search for Extraterrestrial Intelligence. Delta. pp. 55–62. ISBN 0-385-31122-2.
33. Glade, N.; Ballet, P.; Bastien, O. (2012). "A stochastic process approach of the drake equation parameters". International Journal of Astrobiology 11 (2): 103-108. http://arxiv.org/abs/1112.1506
34. http://www.space.com/22648-drake-equation-alien-life-seager.html
35. https://www.cfa.harvard.edu/~ejchaisson/cosmic_evolution/docs/fr_1/fr_1_site_summary.html

36. TESS: Transiting Exoplanet Survey
     Satellitehttp://space.mit.edu/TESS/TESS/TESS_Overview.html
37. Explore James Webb Space Telescope, NASA: http://jwst.nasa.gov/facts.html
38. http://cr.middlebury.edu/biology/labbook/diffusion/frap/membranes/chap1.htm
39. The Blind Watchmaker, by Richard Dawkins, pp. 44-50, WW Norton & Company, 1986
     New Your, London

Chapter 10: The Future of Man

1. Richard Dawkins Net Worth: http://www.therichest.com/celebnetworth/celeb/authors/richard-dawkins-net-worth/

Index

Image Credits

The purpose of *Evo-illusion of Man* is purely to challenge the accepted modern scientific norms regarding the origin of Homo sapiens. *Evo-illusion of Man* is a not-for-profit work, a scientific challenge, and educational in nature. I am therefore claiming *fair use* in my use of all photographs, diagrams, and drawings. Of course, if fair use were not in existence, scientific challenges could never be made. Anyone could say or write any scientific proposal they wanted, and if that was accepted by a large number of people, and not to be challenged, science could never advanced. According to 17 U. S. C./107, in determining whether the use made of a work in any particular case is a *fair use*, the factors to be considered shall include whether the use is of commercial nature, or for nonprofit educational purposes; the amount of the work utilized, in the case of written documents; the nature of the work; and the effect the use will have upon the potential market or value of the work. Evo-illusion is of educational and research nature, and it will have no effect whatsoever on the value or potential profits of any photographs, drawings, or diagrams. Most photos and diagrams come from taxpayer-funded institutions, such as the National Aeronautics and Space Administration, and the Smithsonian National Museum of Natural History. Taxpayer funded documents and images are the property of the taxpayers. As such, they are available for use by those taxpayers. Many images in this book are in the *public domain*, and therefore can be freely used under any circumstances.